PES TS

TO PEACE

ilitary-grade
est control
ecrets

ARK MOSELEY

Rethink

First published in Great Britain in 2023 by Rethink Press
(www.rethinkpress.com)

Cover image licensed by Ingram Image

Contents

This book is dedicated to all the tradespeople who work tirelessly through the muck and sweat to keep the UK operational. Thanks!

Foreword

It is rare to meet someone who has done so much in such a short space of time. The challenge of knowing what you want to do is something many young people struggle with when leaving school, but Mark embraced the army and learned a trade – and a useful one at that. Electricians will always be able to put food on the table! Having served Queen and country and learned the life skills needed to work as part of a team, at times under extreme pressure, he embarked on the next stage of his life. Many people enjoy travel, but few immerse themselves in the real experience. The military teaches people to endure considerable hardship for the occasions when they may have no alternative. When you know your limits, it's easier to live within them.

Mark sought out the extremes, places with no electricity and then those with even fewer creature comforts. 'Easy' has less appeal for him than 'challenging', and after a year or so of travelling, Mark found himself back in hostile territory, using both his engineering and military skills to perform an essential security role – but this time as a civilian contractor.

Eventually the world of business beckoned, and after a short period working for someone else, Mark did what was (to him) the only logical thing and started PestGone Environmental. Like many people starting their own businesses, he believed he could do a better job than most of the industry.

Many small- and medium-sized businesses (SMEs) are run by owner-managers or founders. Some become little more than one-person businesses; others, like Mark's, take on employees, offer employment opportunities and look to scale up. This is the most perilous part of starting a business and the bit that puts many people off. If you are doing it on your own, it can be incredibly lonely, and most of the time you are going where you have never been before. Comfort zones cease to exist, and it can feel like the world is on your shoulders. Legislation considerations have become part of every small business, and despite successive government promises to deregulate, the issues around compliance grow every year.

Mark contacted Heropreneurs, a small charity that supports military veterans and serving personnel's spouses that want to set up and run their own businesses. I started working with Heropreneurs at the end of 2018 after selling my business and the charity connected me to Mark.

Making the decision to start a business is the easy part, although it may not feel like it at the time! Getting through the first few years puts you in that exclusive group of companies that have survived, which is no small achievement. Employing people is probably the biggest challenge for most businesses. Good people are in short supply. Mark's experience in working with diverse groups of people in the military has stood him in good stead. His story makes interesting reading, with good tips included along the way. A whole new chapter is opening up with Mark's exposure to TV. Good luck!

Jon Herbie
Heropreneurs Business Mentor

Introduction

The creatures we call pests can have a detrimental effect on our physical and mental health, whether they are rats peeing around our homes or workplaces, or bedbugs sucking our blood while we sleep. Like us, pests are all fighting to survive in this often-cruel world, and they have been around for a lot longer than we have.

The bubonic plague, sometimes referred to as the Black Death, killed 30–40% of the population of England back in the fourteenth century and an estimated 78 million people worldwide before it reared its head again in the seventeenth century. The plague was caused by rats that carried the disease.[1] Fleas fed off the rats' blood and passed the plague on to humans via their bites. Both pests are still thriving in our modern world, but fortunately, the Black Death has pretty much been eradicated across the planet.

According to the World Tourism Organization (UNWTO), in 2018 there were a record 1.4 billion international tourist arrivals, and the International Chamber of Shipping estimates that some 11 billion tons of goods are transported by ship each year.[2]

The danger of globalisation is that there is the potential for international distribution and establishment of harmful organisms in different regions. Incidents of infectious diseases appear to be increasing because, due to the movement of huge numbers of people and goods, pests such as rats are infiltrating all continents.

In the UK, we have a whole host of pests which often have several subspecies, for instance the garden ant, pharaoh ant, brown ant and ghost ant. My company, PestGone Environmental Ltd, treats more than eighteen different pests in the UK and all must be respected. Some of the cleverest creatures on earth are, in fact, pests. Rats have an incredible navigational brain to operate effectively in pitch-black locations such as the sewers because they have adapted as a species to their environment. The rat also has the second largest number of smell genes on the planet, just being pipped to the post by the African elephant.[3]

There are occasions where we humans can work with rather than against pests. The spider, for example, fascinates scientists. An article by the University of Wurzburg in 2019 entitled 'New research finally explains why spider silk is so incredibly tough' discusses how strong a spider's web is in comparison to its sheer weight and explains that scientists are adapting their technologies based on a spider's silk.[4] Pigeons, often considered vermin, have saved hundreds of thousands of lives over history as they passed messages across enemy lines during wars. They were also carried onboard Naval ships so that, in the event of a U-boat attack, the bird would be released with a note for HQ detailing the sinking ship's location. And today, rats are being used to sniff out landmines.

Whatever you think of pests, and however disgusting you believe them to be, know that they are not stupid. Any species that is thriving in today's society is a hardened, adaptable creature that should be respected.

Why do properties need pest control?

I have written this book for all property owners, whether it be a domestic home or a commercial building. Pests are no different to us when it comes to the basic living requirements: they can and will infiltrate anywhere that has safe shelter, a readily available food source and moisture.

'Speed is the key, Moseley,' my former Army instructor would shout at me as we were manoeuvring across the English countryside, attacking a mock enemy position, and this motto applies perfectly to pest control too. Identifying the pest, the source (entry route) and whatever is attracting it (food) and gaining control as quickly as possible is crucial.

I fully understand that many business owners in the hospitality sector see pest control as an evil necessity. They feel they are throwing good money away, simply to tick a box to keep the health inspector happy. Some, including high-profile chefs, have refused to pay pest-control companies after months of treatments at their restaurant, believing the task to be menial. They believe that as their property no longer has any pests, they no longer require pest control. However, what they don't realise is that the reason they do not have any pests is because pest control is in place. That is what is nipping the problem in the bud in a timely fashion before a more serious infestation takes hold.

I have seen first-hand what can happen when a business owner decides to discontinue pest control. A bakery lost everything overnight, receiving tens of thousands of pounds in fines because mice droppings were found in their hot cross buns. The owner had not heeded the warnings of his staff and refused to pay for continued pest-control services, partly because finances were tight and partly because he believed he could control the rodent problem by himself.

Most businesses, especially those in the hospitality sector, require some form of pest control and prevention in place. The hospitality sector is heavy with legislation and business owners must document their pest-control approach in their method statement, the pesticides they use in the Control of Substances Hazardous to Heath (COSHH) assessment and any dangers to their staff or the public in their risk assessment. All reports and assessments must be logged and kept onsite for the visiting environmental health authorities, who may stop by unannounced to investigate the hygiene of the property at any time.

If you own a hospitality business, having all this information at hand will save you time, stress and possible financial penalties, or worse, forced closure, but it can be complicated paperwork if you aren't aware of the legislation and are not an expert in pest control, and you risk being fined if it is not completed appropriately. Investing some money now in engaging an expert to keep on top of your pest control, including all the paperwork, can save you a lot of expenditure and heartache in the future. It may even save your business.

Daily, I see people struggling to eradicate pests from their homes and workplaces. It affects their physical and mental health, which results in their work and personal life becoming unbalanced, sometimes even to the point where relationships break down. One of my clients had to move out of their home for a month and live in a hotel because they'd had a mouse co-habiting with them and physically couldn't bear it. Another threw away thousands of pounds' worth of clothing and shoes due to a couple of the garments having clothes moths. These may seem to be extreme cases, but a pest can have ramifications – personal, financial and wellbeing. All over something that could so easily have been avoided if the person remained calm in a somewhat chaotic situation. Of course, that may be easier said than done.

The 6Cs

Before I introduce my six pest-control principles, allow me to share a few details about my background. My name is Mark Moseley and in 2017, I became the founder of PestGone Environmental Ltd, a London-based pest-control company. Before I became a pest-control expert, I was a soldier and security specialist operating in remote, hostile environments around the world. In 2016, after serving in Afghanistan, Syria, Iraq, Israel, Jordan, Ethiopia and countless other countries, I found myself in London, falling into the pest-control industry. As well as working in security, I am also an experienced electrician, but working on construction sites didn't give me the sense of purpose that I craved.

In this book, I will detail how I created the 6C pest principles which I have adapted from the 5C immediate action drill taught in the military on finding an improvised explosive device (IED): confirm, clear, cordon, control, check. The 6C principles, when they're applied effectively, will help you to eradicate unwanted visitors – pests – from your home and/or business.

Whether you are a property owner or the facility manager at a sports stadium, hotel, night club, restaurant or any other building, this book will enable you to identify, control and eliminate pests for the long term. As I lead you through the 6C pest principles – confirm, clear/clean, cordon/contain, control, check, communicate – I want to make the book more engaging by telling a story of my personal experience of using each C, either on operations in the military or when I was on my travels. Each chapter will also contain a case study to show how the 6Cs work in the world of pest control. My aim is to highlight that remaining calm and in control of the situation and following these principles, you can achieve the right result. As my old Army instructor told me:

Speed is the key.

Always remember this.

ONE

Of Mice And Men

Pest control is seen by many as a derogatory trade that attracts only people from uneducated or lower-class backgrounds. I disagree. I work with some of the cleverest, most hardworking individuals that I have ever met. The industry is recession proof, pandemic proof and here to stay for the long term, because pests are resilient and adaptable creatures. The biggest name in pest control, Rentokil, has been operating since 1924, when it was founded by an entomologist trying to eradicate death-watch beetles in an old hospital.[5] In his article 'New list: Top 50 best restaurants in London 2022', James Bellis says that there are over 20,000 restaurants, 1,500 hotels and around 9 million people in the UK's capital city alone, so it is no wonder pests are thriving with the massive amount of food waste they must generate.[6]

One of my favourite sayings comes from Walt Disney: 'I only hope that we don't lose sight of one thing; that it was all started by a mouse.'[7] Walt was, of course, referring to Mickey Mouse and how that one chirpy, talking mouse created a huge magical empire, but I like to use this line in my industry – especially when I see a well-built 100 kg man standing on his mattress, wielding a cricket bat because he has seen a mouse run under his bed. I fully understand that people react differently to pests, and that the man was extremely frightened because he suffers from musophobia, the fear of mice. In fact, Walt Disney himself suffered from musophobia as a child,[8] so to stop other kids experiencing the same fate, he created Mickey, humanising the animal. Sadly, Mickey didn't work for my cricketing customer.

I will be leading you through my 6C pest principles, which I will introduce in the next section, throughout the chapters of the book, but first, I want to touch upon how I created these by using the 5C military reaction principles when locating an IED. Just like IEDs, pests can have a detrimental effect on our physical and mental health, so we need to apply care when we're dealing with them.

The 5Cs are:

1. **Confirm** the device is an actual IED. In the past, people have panicked after coming across an unidentified item while digging their garden and called for the bomb disposal team, who

have arrived only to unearth a lawn roller. If, however, the item is an IED, the team will apply the next four Cs.

2. **Clear** the area of people. This is to ensure human life is preserved in the event of the device exploding.
3. **Cordon** off the area. The last thing the bomb disposal team needs is to have people and vehicles coming through a dangerous site, putting their and others' lives at risk.
4. **Control** the cordon's logistics. Who is allowed into the area and who is not?
5. **Check** for secondary devices. Often, terrorists plant an obvious device with the aim of drawing in the emergency services to a scene, only for an even bigger hidden device to kill key people like the bomb disposal team and other first responders.

These 5Cs, which served me so well during my military career, now form the basis of everything I do in my pest-control business. I have added an extra C for communicate, which completes the set of principles needed to control a pest infestation. By communicating effectively, I keep people safe and ensure they understand what they need to do to prevent a re-infestation.

Let's now look at the 6C pest principles in more detail and see how they compare to the 5Cs of the military.

The 6C pest principles

The 6Cs, on first glance, look similar to the military 5Cs. There's a reason for that – they are! In this section, we're going to look at them from the perspective of pest control rather than bomb disposal.

1. **Confirm** you really have a current pest problem. If you have seen a pest, then that's your confirmation, but my company gets many calls from customers who have mistaken spilled foods such as bread seeds and peppercorns for mice droppings, or a torn carrier bag or piece of clothing to be rodent or insect damage. In one case, a structural surveyor found a 20 mm hole where water pipes used to run but believed the hole to be historic woodworm damage. If woodworm created 20-mm-diameter holes, then buildings would be dropping like flies!
2. **Clean/clear** away any debris you find, even if you are unsure if a pest is present. If you have confirmed you have a pest, the next steps you take will depend on which pest you have. I advise you bleach all kitchen areas and clear away all accessible foods. Vacuum carpets if moths or fleas are present. Wash clothing and/or bedding if you've seen textile-feeding pests such as moths or parasites such as bedbugs.
3. **Cordon/contain** – put all exposed foods into sealed containers. This includes breads, cereals,

rice, spices and any other stored foods which are easily accessible for pests. With textile-feeding pests, ensure all wool or silk items are quarantined in zipped bags, or at the minimum those that are tied at the top. Wash or freeze them as soon as possible.

4. **Control.** The treatment you use to control an infestation will depend on the type of pest you have found. Pesticides and/or traps will be crucial to gaining control quickly.
5. **Check.** After you have carried out the relevant pest control, make regular checks to ensure the problem is gone… and stays gone.
6. **Communicate** the problem and the steps you have taken to resolve it with the people you are living or working with. This may involve speaking to your neighbours about the infestation as it is important to get on top of pest problems quickly and remain pest free for the long term.

We will cover each of these principles in the chapters that follow, going into detail on the types of pests you're likely to encounter and how to treat them. As this case study shows, though, there are some occasions where the 6Cs simply won't be enough. That's when even experts like the PestGone team must hand the problem over.

CASE STUDY: TOO BIG FOR THE 6CS

My company has worked with the Port of London Authority for a number of years to collect dead animals which have washed up on the banks of the River Thames. The PestGone team collects swans, cats, seals and porpoises throughout the year. We've even collected dogs with lacerations around their faces and necks, mainly Staffordshire bull terriers that we realised were likely being killed in a fight and then tossed into the water. It is sad to think that dog fighting still goes on in the UK for people's entertainment.

One day, my team and I were called out to collect a porpoise, but when we arrived, I noticed the animal was a good 4 m long. Usually, porpoises only reach about 2 m in length. It also looked to weigh a tonne, if not more.

It turned out that the floating corpse was a juvenile minke whale. If it was left exposed out of water for too long, then gases inside the animal may result in it exploding if it was shaken about. Not that our van could have taken the weight of the whale on its roof, but it did cross my mind that the story would have hit the news headlines if it had exploded while we were driving through the Blackwall tunnel. I guess it would have been a great bit of publicity for the business, at the very least.

With the animals we do collect from the river, we take them to our freezer storage unit, from where they are collected later and incinerated, but in the case of the minke whale, we never did get our moment in the spotlight. Due to the animal's sheer size and weight, a specialised company had to remove it in a large metallic shipping container.

Summary

In this chapter, we have identified how the military 5Cs for IEDs easily transfer to the world of pest control. With the addition of communicate to create the 6Cs of pest control, you can see how all these principles, when implemented fully, will help you gain control of your infestation problem and reduce the risk of the pest returning in the future.

I truly believe that if you are dealing with any pest, great or small, by following these steps and aligning them with the pest problem you are facing, you will win the battle. But if that battle proves too much for you to resolve alone, there are always experts out there who can deal with it for you.

Good luck!

TWO
Confirm

In this chapter, we will look at the importance of confirmation, both in a military setting and when working with pests. In any military situation, we need to confirm what we are actually dealing with. This allows us to plan how we will tackle the problem. In a pest infestation situation, confirmation of what type of pest we are dealing with is essential to know how to treat the problem so that the pest is eradicated fully from our home or workplace.

Military confirmation

In November 2002, I was a serving soldier with the Royal Engineers, stationed in Antrim, Northern Ireland. I had been in the country for two months of a six-month operational tour from my parent base of Hamelin, Germany. News of the forthcoming Iraq war was becoming the popular talking point then, but a more urgent task came through that month, which was that many British soldiers in the UK would be covering the national firefighters' strike. UK firefighters were in a dispute with the government over pay.

Operation Fresco was the name given to our task of taking over as firefighters. We would do this for forty-eight hours at first, but would continue in the role if an agreement was not established between the firefighters and the government. The soldiers on the UK mainland had been given green fire vehicles known as Green Goddesses, but due to the political situation in Northern Ireland, green was not a popular choice of colour as it heavily represented the Army. Instead, the powers-that-be decided on the colour yellow. Not a subtle, calming yellow, either, but the brightest, most illuminating shade you could imagine. Consequently, we were the proud recipients of Yellow Goddesses.

After we had been given some training on using the firefighting equipment and how to cut people out of vehicles in the case of a road traffic collision, sixty personnel from my squadron were ready to hit the ground running to help in the firefighting efforts. The cold, wet Northern Irish winter was in full swing as we were assigned to the county of Derry, which was infamous for 'The Troubles' in previous decades. My role was to be the water hydrant operator, which meant, upon arriving at the scene of a fire, I would help locate and lift the fire hydrant cover to connect the pipework and hoses and turn the water on. It was a simple job – so long as the communication channels were clear about when to turn the water on and off.

In Northern Ireland, we had to be more security conscious about what we were doing than our colleagues on the mainland, especially about where we were standing and what we were lifting. The Police Service of Northern Ireland (PSNI) would escort us to the scene of the fire, where they would help by holding back any hostile crowds forming, but an intelligence threat had come in indicating that IEDs might be put under the fire hydrant plates. When we – and by 'we', I mean I – lifted the steel plate, I could lose my limbs or even my life in the event of a blast. To counter this, I was provided with a device that had a pull cord of about 40 m in length, which was connected to the hydrant's steel cover plate so I could pull it up from a distance to ensure we were not injured if a homemade bomb had been placed under the plate.

The pulling device was poor at best. On occasion, it failed by simply falling over sideways when I pulled the draw cord, slowing down the firefighting efforts. I would check the seals around the hydrant plate to see if they looked disturbed or whether the gap where the plate sat in its square housing for long periods had mud or debris around it. If there was undisturbed mud and mess in the outer gap, I would lie flat on the ground and slowly lift the cover plate lid with a crowbar, checking for any trip wires.

We successfully covered the first midweek fire strikes, but there would be different challenges on the next as it fell over a weekend. On the Friday night, as we were driving through the Derry nightlife at 10pm in a fire truck that was brighter than the sun, we heard BANG, CRASH, SMASH, as bottles, coins, stones and any other objects partygoers could get their hands on hit the sides of the vehicle. Most of the people I encountered during my time in Northern Ireland were among the friendliest and warmest I have ever known, but there were those who didn't appreciate our presence in their country.

This night, we were heading to a domestic house fire in an infamous area of Derry called the Bogside. We had been warned that a large crowd had formed and the area was likely to be hostile, so six armoured police vehicles were accompanying our two yellow fire trucks. We roared into the area at an eye-watering 30 mph – the top speed the Goddesses could muster – behind the police vehicles. The crowd parted to reveal a terraced house ablaze.

I unlocked the steel slide-shutter door and jumped from the truck with my crowbar and the pipework to connect to the hydrant once I'd located it. One of the crowd of locals directed me to the hydrant, which immediately made me suspicious. On the ground next to it were cut-off insulated blue wires – why a bomb maker would need to cut wires at the scene while installing an IED, I didn't know. It didn't make much sense, but I started to worry just the same.

A group of teenage lads were also hovering, shouting at me to get the water on quickly in case there were people in the house. As always, speed was the key, so I had to act fast. I checked the muddy seal around the hydrant plate housing as well as I could with a torch as I didn't want to use the pulling device after the failures I'd had previously, especially in front of a crowd of 100 or so people, which would make us look unprofessional and would, no doubt, bring about a backlash in negative publicity.

The seal looked undisturbed. If there had been an IED underneath the hydrant, the likelihood is that the locals would have been warned because no terrorist would want to harm their own citizens, especially the youngsters who were around us, but I was still nervous. I warned the police escort and the crowd to give me a bit of room as I lay on the wet, cold ground to lift the hydrant cover slowly by 5 cm to confirm that there was not a bomb below it that could kill or maim countless civilians and – more importantly, to me anyway – me! My helmet with visor attached offered minimal protection at best, but nonetheless, I steadily lifted the cover as my teammate shone a torch in the gap in the hope of catching a reflection of a wire. Fortunately, there was nothing.

'All clear!' I shouted. My teammate and I removed the plate and connected the pipework and hoses, getting water on to the fire immediately. The search teams wearing oxygen masks entered the home to confirm that there were no people or pets inside. While this was going on, my teammate and I stood with one PSNI officer at the water hydrant about 80 m away from the main action. The crowd abused us all the more for supposedly being bystanders at this point.

Twenty minutes later, seemingly out of nowhere, a deluge of bottles and rocks landed between us and the team dealing with the house fire. Then a large glass bottle smashed and spread flames across the road. My teammate ran back to attend the fire and the standby police, dressed in riot gear, leapt from the back of their armoured vehicles, segregating the crowd from the team at the house. This left me and my police escort alone at the fire hydrant.

When the fire was out, I turned the water off, disconnected the pipework and sprinted back to the Yellow Goddess, which was covered in dents and scars from the scenic trip through the town centre an hour earlier. I helped my colleagues as they quickly rolled up the hoses while more projectiles were thrown. The police were now using their batons to deal with the Friday night crowd in the residential street as we stand-in firefighters jumped in the truck and battened down the hatches.

None of us were seriously hurt from the objects being thrown, but bangs and crashes continued to sound against the steel truck. Someone threw half a house brick at the windscreen, but it was protected by a steel mesh, so the brick bounced off and hit one of the other rioters on the head, knocking him to the ground. Thankfully, the 1950s engine started first time and we eased around the injured man, lying on the floor and still scratching his head as we left the scene as fast as the truck would carry us in a cloud of black smoke from the exhaust. It was like the sketch in the famous BBC sitcom *Only Fools and Horses* when Del Boy parts the riot outside Nelson Mandela House with his illuminous yellow Capri Ghia.[9] Except the horn on his car worked, unlike ours.

This real-life example of a military situation shows how the first principle of the 6Cs – confirm – played a vital role in keeping me, my colleagues and the public safe. Despite the abuse from the crowd, it was essential for me not to skip this step as the consequences could have been fatal if an IED had been placed in the fire hydrant.

Now let's look at this principle from a pest-control perspective.

Pest confirmation

Taking the time to confirm if you have a pest or not, even under the pressure and stress of knowing something is invading your home, is crucial. It can save damage to your property, whether it is your home or your business, and ultimately save you money.

Many people panic upon moving the sofa during their spring clean and finding what they believe to be mouse droppings underneath. Those droppings may have been there for months, if not longer, so taking the appropriate steps to confirm whether a pest is currently in your property is important. Different pests have different telltale signs, so in this section, I will list the four key signs that confirm if you currently have a pest in your property or not.

Live sighting

If you see a mouse or rat run across the kitchen floor, then that is all the confirmation you require. The same goes with insects. If you see ants physically stealing the jam, bedbugs crawling over the mattress, flies circling overhead or moths feeding on the knitted hat that your nan made you, you have comprehensively identified and confirmed the problem.

Monitors

Sticky insect monitors are great for catching spiders, and insects such as ants, cockroaches, flies, moths and more. Some monitors have a pheromone scent that lures the insect in, such as the moth monitor which gives out a female scent, attracting only the males.

Customers will say, 'But what if I use this monitor and don't have pests, and then one comes through the window attracted by the smell?' This is highly unlikely to happen, so I would always advise you to use these monitors, but only if you believe there is an insect present. Never use glue boards for rodents, though. These are inhumane and, in the UK, legislation has been passed in Parliament ensuring that they can only be used by professional pest-control technicians or other trained personnel.[10] The legislation comes into force in 2024.

Hearing noises

If you hear noises throughout the night, particularly scratching, then it is highly likely you have a rodent in your ceiling or wall cavity. Mice and rats are predominantly nocturnal and gnaw to file down their teeth. A rat's gnawing is incredibly loud.

If you hear footfalls, it is likely to be a larger animal such as a rat or squirrel, not mice. I call squirrels 'the dusk and dawn' animal. They spend most of their time outside during the day and usually sleep through the night, unlike mice and rats who have the opposite sleep cycle. If you hear noises at daybreak and/or at the day's end, especially in the loft space, it will likely be squirrels.

Squirrels will rarely live between floorboards or in walls as, being bigger than mice and used to open environments, they do not like tight spaces and struggle to move in cavities. That said, they may sometimes nest in the boxed soffit part of the roof where the rainwater gutters are fixed, if your property has this feature.

Birds nesting in the loft space are protected by the Wildlife and Countryside Act in the UK, meaning they cannot be removed without special notices being issued.[11] The same goes for bats, which are even more heavily protected.

Some insects make a soft tapping or scratching noise in the loft ceiling. It may be wasps using their mandibles to strip wood shavings from the joists, which they mix with their saliva to make their nest. I once came a cropper to wasps when drilling an opening to enter through a plasterboard panel in the eaves of a house. The customer had said he'd heard scratching and a 'soft running noise', so I decided I would install rodenticides (poisons) as, considering the shallowness of space in the roof void, I believed it was mice or rats.

As soon as I drilled into the cavity, worker wasps flew out, giving me the good news with their stingers to protect the nest. Sadly, I am allergic to wasp stings and with the fifteen or so I received from my attackers, I blew up like a frog. That incident taught me a valuable lesson: not to be blinded by what the customer tells me about the noises they hear, which of course comes back to the importance of confirming the exact pest I'm dealing with.

Another insect I have known to be found, particularly in old buildings, is the death-watch beetle, which starts off as a woodworm. The adult beetles make a distinctive tapping sound, which is their mating call.

Bites

This can be tricky as a key sign of pests as some people will show bites and some won't, and not all bites come from a pest in the home or workplace. In summertime, insects are active with the warmer weather, which increases their metamorphic life cycle. This means that you may suffer a bite while out and about.

Bedbugs are also more prevalent during the summer as they may be brought back in your suitcase after a holiday. An adult bedbug can grow to the size of a little fingernail and is easily identifiable. They usually bite the upper torso as this is the part of the body that is exposed outside the bed covers when you sleep, as well as being the area closest to their likely nesting spot – the headboard.

A point to note when you're checking for bedbugs is that they will bite three or four times in a cluster or a row. They usually hide around the back of the bed's headboard or inside any creases where the lining fabric may be stapled on. Another favourite nesting spot is around the wooden slats of the bed, so always check here. They do not like metallic bedframes, but check the wooden areas of the bed, furniture close to the bed and behind framed pictures above the bed. If you see black dots, this is their secretions and the tell-tale sign you have bedbugs.

If you have birds nesting in your loft space or a redundant open chimney or flue, then bird mites, which are parasites that feed off the bird's blood, may migrate into your property and begin feeding on you too. In large numbers, these mites are just about visible to the naked eye.

Mosquitoes and midges are common biting pests. In the warmer summer months when people sleep with their windows open, these insects enter and bite. I get a lot of calls in the summer from customers believing they have bedbugs, when it is more than likely a flying insect coming in through the window.

Some spiders are also biters, so be mindful of any arachnids that may be close by.

Things to note

- If you suspect you've found rodent droppings, vacuum them up immediately for health and safety reasons. This will also reveal whether they are old or new droppings.
- If you're bitten, think about what time of year it is. If it's summer, then more biting insects will be present, so don't automatically conclude you

have bedbugs, especially if no one in the home has been away on holiday and there are no pets present to transmit fleas.

- Record noises if you hear them, saving the exact time of day or night on the recording. You can then use this as part of your investigation into whether it is a pest or just the boiler shutting down in the evening.

CASE STUDY: WHO'S THAT TAPPING ON MY CEILING?

I met a landlord at his property because his tenant was hearing noises first thing in the morning and last thing in the evening. This immediately indicated to me that it would likely be squirrels.

Upon inspection of the property, I discovered that the landlord had built into the roof space of the house. If the pest was a squirrel, it would have struggled to fit between the roof's external tiles, the insulation and the plasterboard ceiling as it was such a shallow space. The landlord asked if I could cut into the ceiling to find out what was going on. This didn't make any sense as it would be destructive to the roof, so I decided to investigate further first.

The tenant was incredibly stressed, believing animals were living above her bed. She explained that the noises were not scratching or sawing sounds, but more of a tapping. I wondered if this was the species of woodworm called the death-watch beetle, but as the building was too new, it was more likely to be wasps. I inspected the outside for any flying insect activity and checked the soffit and the guttering.

As I moved the outside guttering next to the bedroom where the noises were being heard, the window opened and the tenant said, 'That was the noise. Whatever you just did then, that was it.' We soon discovered that the plastic guttering was expanding in the heat of the morning, and then contracting in the evening as the day cooled. This caused a tapping sound, which was what the tenant could hear.

I secured the gutter with cable ties to prevent it moving with the temperature change and this was the job complete. If I had not used my experience and confirmed my theory, it would have cost the landlord hundreds of pounds in fees for a plasterer to repair the holes I'd have created in the ceiling.

This case study emphasises the importance of confirming you actually have a pest. Make sure you book a professional with a wealth of knowledge and experience to identify any abnormalities you hear or see.

Summary

When my military colleagues and I were dealing with the house fire in Northern Ireland in the story at the beginning of the chapter, confirmation was key to the safety of the surrounding properties, the locals and ourselves. This incident was one of many in my time in Northern Ireland when we had to confirm what we were dealing with to ensure our safety, but I would not have changed it for the world.

Similarly, confirmation of pests in your home or business environment is crucial to save you time, stress and money. This is why 'confirm' is the first of the 6Cs of pest control. If you fail to confirm, you are confirming that you will fail.

If the noises you are hearing in your property or the bites on your body are the work of a pest, then act quickly. Otherwise, the consequences can be very costly. If you are worried or struggling to identify what you may be dealing with, then get a professional in to help you. Spending some money up front can save you from having to spend out thousands in the long run.

THREE

Clear And Clean

In this chapter, we will look at the importance of the second principle of pest control: clean and clear. We will delve into the specifics of how to deal with food waste, which is the draw for most pests, and examine key tips when you're cleaning away food that will help you reduce the risk of pests invading your property.

Let's look first at the importance of this principle in the military world, where it is simply called clearance.

Military clearance

My military regiment, the Royal Engineers, had been sent to help protect the residences in Cluan Place, a cul-de-sac in Belfast. It was a working-class Protestant area in the inner city, surrounded by Catholic Republicans. In recent months, there had been violence between both sides, known as the 2002 Short Strand clashes. Five civilians had been shot, twenty-eight PSNI officers injured and several families forced from their homes. It was the Queen's Jubilee, and the Protestants were being accused of draping red-white-blue Unionist bunting on the rails of St Matthew's church in Short Strand.

We were tasked with fencing the outer perimeter of where the Protestant houses backed on to Republican properties. When we arrived on site on a weekend in civilian clothing in unmarked vehicles to offload the stores and set the site up, a group of young men had gathered at the entrance of the cul-de-sac. A firework was set off in our direction and we had to take cover behind our vans and call for assistance.

The police arrived and things calmed down. The last thing anyone needed was the violence to rear its head again, so we decided that the weekend was not the best time to carry out the work. We returned in the week in military clothing and military vehicles and set about installing the high fencing around the perimeter of the cul-de-sac.

I happened to be 6 m high, hanging from a harness on one of the steel fence posts, when a ladder appeared from a neighbouring garden on the Republican side against the steel fence we had just installed. A man in his fifties appeared, smoking a roll up.

'What the f*ck do you think you're doing?' he said.

Even though it was obvious, I knew not to make a smart remark for fear of things turning aggressive.

'Putting up this fence as we heard there's been a bit of trouble here recently. Hopefully it may help calm any tensions. I couldn't nick a roll up off you, could I?' I didn't smoke roll ups, but I wanted to diffuse any anger he may have felt towards me and try to find some common ground with the man. It could have gone either way, but fortunately, he passed me one through the grated fence and we had a civilised chat, him on his ladder and me swinging from my harness.

Ten minutes later, we ended the conversation if not as friends, I suppose you could say acquaintances. I even joked that he could bring me a cup of tea the next time he made an appearance. He told me to piss off, but at least I tried.

An hour later, a man was reported walking the streets with a crossbow, cursing the Army. The Royal Irish regiment managed to find him and speak to him. He was licenced to have the weapon and would only be arrested if he used it to endanger life. This didn't reassure us too much.

After two days, the fence was complete. Daylight was fading and our small storage ISO shipping container, positioned at the entrance of the cul-de-sac against a free-standing wall, was being moved when an IED was discovered. It was a pipe bomb about 2.5 ft long with plastic explosive compacted into a steel tube. It was likely that the safety cord, which burns extremely slowly, had been ignited. A detonator would normally be pressed into the material, but when the item had been thrown, we assumed it must have fallen out of the plastic explosive, which resulted in the device not triggering and prevented the plastic explosive going off inside the tube. Nevertheless, this bomb was still dangerous. If it did go bang, parts of the pipe would act as deadly steel shrapnel, which could easily maim or kill someone.

We called military explosive ordnance disposal specialists, along with the PSNI. Then we cleared the area of civilian personnel, evacuating people from their homes in case the device did detonate. Once the area was completely clear of people, the specialist teams began working to remove the device safely. We were told to head back to our barracks in Antrim as we were no longer required. On returning to the camp, I switched on my TV and the Northern Ireland national news was covering the pipe bomb incident as one of its main stories.

Remaining calm in a pressured and hostile military situation is important. In this case, it eased tensions with the gentleman on the ladder and made sure the IED was dealt with in a safe and timely manner, minimising the risk to human life. The same sense of calm is useful when you discover an active pest infestation. I was able to diffuse the situation by creating a common ground of smoking the same sort of cigarette as the neighbour. Creating a pest plan and calming yourself and or others in the property is important to gain control quickly, and the first step on that plan must be clean and clear.

The best way to create calm in a pest situation, when your instinct may be to panic, is to clear and clean the area in which you have identified a pest infestation. Clear away any foodstuffs and dispose of those that could be contaminated by pests, which might seriously harm your health. Getting rid of food sources is one of the main steps in eradicating pests.

We will now cover the key things to ensure you are clearing and cleaning from your property, not only to keep you and others safe, but also to eradicate whatever it is that is attracting pests to your premises.

Pest clearance

Clean, clean, clean! When it comes to the majority of pests, cleaning away whatever is likely to be attracting them is the quickest way to deter them and, in some cases, eradicate them. If you remove the source of attraction, often food, the pest will have no choice but to scavenge in other places to feed.

Food

If you leave sugary foodstuffs like jams in open containers or spill sweet fluids like orange juice on the floor or counter, you are likely to draw in insects such as ants during the warmer months. When one scout ant finds a food source, it will run back to the nest, leaving a pheromone trail in its wake to signal all the other ants to follow it to the jackpot of sugary delights. If you observe ants, they may sometimes look to be kissing. What the scout ant is actually doing is passing food on to others going the opposite way on its trail.

Rodents are scavengers and will home in on an easy meal. The rat has one of the best noses in the animal kingdom, with the mouse not far behind. If you have a pet and leave the remnants of cat food, dog food, rabbit food, etc out, then you are asking for trouble.

'But I have cats, they eat mice,' you may be saying. I wish I had a pound coin for every time I have heard that line. Remember the old saying, 'While the cat's away, the mice will play'? It's true. Mice may set their sleep patterns for the optimal time to go and raid the cat's bowl, usually at night. If the cat is out all night, the mice can sense and smell this, and I have known a number of cats to run away from rats who are feeding from their bowls.

The same goes for dogs. On one occasion, a client explained to me that his dog had suddenly stopped drinking its water. I found a mouse dropping in the water, so it was likely that mice were bathing in the shallow dish and possibly urinating in it. Once the mouse infestation was treated successfully, the dog found its taste for water again.

The clear lesson from this is if your pet is not feeding from its bowl, then stop leaving its food, even the residue-stained bowl, on the floor. Put the bowls up high on a table or on the kitchen units, or better still, put them in an enclosed container.

Be mindful that your fridge freezer cable is not plugged into a socket above the kitchen units as mice will climb it to gain access to the worktops. The same goes for gas or heating boiler pipes or gaps around the top of a poorly fitted oven or washing machine. These holes can be blocked or the pipe work boxed in with an aesthetic cover. If your furry intruder gains access to the kitchen units, then you're in trouble; you may find droppings in the toaster, under the microwave, around the condiments and even in the bread bin.

Never leave bags of pet or bird food in basements or under the stairs. This will attract rodents and insects who will think they have an easy, reliable and abundant food source, so will nest locally. These areas are likely to be where the building's cable and pipe services enter the premises, and rodents will climb these to access your property through the gaps around them.

Sheds and garages are another prime location for pests due to the storage of foods such as bird and grass seeds. My father, a keen fisherman, left fishing bait in his shed for some months. As a result, mice not only set up home there, but they destroyed some of his fishing bags and rod handles with their excessive gnawing. This resulted in my mother's cat, who was happily asleep on the sofa, being picked up and thrown in the wooden shed for half an hour to earn its keep. Despite being unhappy at having been woken from one of its several daily naps, it did catch a few mice, with the rest fleeing for their lives. This cat was a mouser, so a contained environment made for the perfect hunting ground.

Food packets that have been opened and left in cupboards or on shelves are prime scavenger treats. Cereals, pastas, dried fruits, powders such as flour, rice and countless others act as ideal breeding grounds for moths or beetles that specifically target pantries and other food storage cupboards if they're not cleaned regularly. Clear out all opened food packages and clean anything that has been spilt. For any open foods that you don't foresee consuming in the next month or two, I would suggest placing these in sealed containers or clip-tying the top to ensure no gaps are left open through which pests can gain access. Empty your food cupboards every six months minimum to make sure you don't have any open stored food that you have forgotten about. If you don't, I guarantee that you will have a stored food pest that will require treating within a year or two. Prevention is better than cure, as I'm sure you know.

By food, I mean anything edible. Rodents will feed off washing powders, soaps, even toothpaste because they can smell that they are organic. Wool or silk garments are food for several pests such as textile moths, often referred to as clothes or carpet moths. The larva stage of complete metamorphosis (egg, larva, pupa, adult) is the one causing all the damage, since it has mouth parts to feed, unlike the adult flying moth. Think of the caterpillar and butterfly; it's the caterpillar creating the holes. Adult moths don't have the mouth parts to feed off garments, so if you see flying moths in your property, expect there to be damage they have caused as larvae to your garments or carpets somewhere close by.

Gardens will attract all sorts of animals and insects. Clear away overgrowth to prevent rats living among the foliage. If you don't, then foxes may stay close by or even burrow in the overgrowth to catch an easy rat. With the entry of one pest, another pest higher up the food chain will often follow. Do not leave bin bags on the ground as mice, rats and foxes may go through them for tasty treats. Bags must be placed in hardened containers with a lid.

Hoarding is a common problem when it comes to attracting pests. We will look at the perils of living next door to a hoarder in the case study at the end of this chapter.

Guano

Bird guano (poo) can cause respiratory health problems in humans, especially if pigeons or seagulls have set up a nesting spot on top of a building where the heating and ventilation systems sit on the roof. Particles in the bird poo may be sucked back in the system and released into people's working or living space. Pigeons nesting under solar panels can be problematic as the nesting material can, over time, pull on small voltage cables, which may stop them from functioning correctly. Pigeons may have bird mites that migrate into homes and workspaces to feed on humans, even when the birds are removed.

If you have solar roof panels, make sure they're proofed, inspected and cleaned regularly to prevent birds and possibly squirrels gaining access. It is important to stop the resultant problems, including bird debris falling into the property's gutters and causing blockages.

Woodworm

It is notoriously difficult to deter woodworm, which feed on soft wood. It is the larva stage of the wood boring beetle, known as woodworm, that creates all the damage, in this case, pinprick holes in the wood. The larvae turn into small adult beetles which mate with females, who then lay their eggs in and around wood. If you have woodworm in the structure of your property, then you must call a specialist in immediately to gain control early on and ultimately save yourself thousands of pounds.

Things to note

- If you find droppings, vacuum them up immediately for health and safety reasons.
- Clear old carpets, boxes or general clutter from storage cupboards, lofts and basements in homes or commercial buildings to prevent the area becoming a breeding and nesting ground for pests.
- Food waste should not be left in an indoor bin for longer than twenty-four hours. Ensure you place outside food waste in secure hardened bins that vermin, including foxes, cannot access.
- Take old clothing to the charity shop if you're never going to wear it again. Seal seasonal wear in plastic bin liners, vacuum packs or, at the very least, suitcases when it's not being used. This will stop textile pests such as clothes moths infesting the garments.

CASE STUDY: A HOUSEFUL OF RUBBISH

I was called to a semi-detached house where mice from the neighbouring property were gnawing their way through the underside wooden cavity flooring to gain entry under the stairs. The problem had been going on for weeks. I would eradicate tens of mice, but more kept coming.

I knocked on the neighbour's door to discuss the problem as the mice were coming from their house. A middle-aged woman let me into a hallway that was filled to the ceiling with bags of rubbish. She had created a pathway down the middle just about wide enough to sidestep through. The bags rustled with mice moving around inside them as I shuffled by. It was the same on the staircase, and the living room was waist-high in newspapers, shoes, clutter and rubbish. To my amazement, two teenagers struggled past me in the corridor, going off to school as though nothing was amiss. I was utterly baffled.

The property was tenanted, so the following week, I met the landlord on site. He asked if I could just get rid of the mice as the woman wanted to continue living like this and he was happy for her to do so. I was confused and, I confess, a little angry. I told the landlord that I could not get rid of a pest problem in the property, given the vast amount of food waste left exposed for the mice to eat, as this food waste would prevent the mice from eating my rodenticides. Added to that was the fact that the bottom of the back door was broken, allowing new mice to enter the dwelling easily, attracted by the scent of the mice currently in residence and the food debris.

In the end, I offered the landlord and tenant my services to clear all the waste from the property as this was the only way to fully control the situation. The tenant did not want the rubbish removed and the landlord said that his hands were tied. I had to explain to my customer living next door that we were fighting a losing battle and I had exhausted all of my options.

I sealed all the entry points in my customer's house with steel mesh. This did keep the mice from coming into the living space, but my customer could still hear gnawing noises in the evenings.

I received a call from my customer a month later. The next-door neighbour had finally moved out and the landlord had instructed the council to remove all the waste. Only then was the rodent problem finally cleaned and cleared away along with the rubbish.

Summary

The military story at the beginning of this chapter highlighted the importance of clearing people from the vicinity of an unexploded IED to protect life. The principle of clearance, or rather clean and clear, is just as important in an infestation situation, and can also influence your life, as pests feeding from spills and food waste may potentially pass on harmful pathogens to you or your family, not to mention the negative effect the thought of them being in your property has on your mental health.

Therefore, keep your property clean and clear of debris, food and clutter to prevent pests taking up residence with you.

FOUR
Cordon And Contain

In this chapter, we will investigate the importance of cordoning and containing the pest infestation. In the military context, cordon describes a surrounding barrier to prevent access to or escape from an area or building. Containment in the case of pests means stopping the spread. I have used both words as the third of the 6Cs of pest prevention as they relate to containing infestations and cordoning off food sources.

Military cordon and contain

In 2003, my unit had returned from Northern Ireland prematurely to begin training for the upcoming move into Iraq. After two months of training in Germany, we were ready to be deployed to warmer, sandier climates, but a week before we were to depart for the holding base in Kuwait, half of the squadron was stood down. The other half would still go and carry out search duties, looking for munitions in Iraq, while the rest, including me, remained behind.

This was a blow to us because – not wanting to sound gung-ho – but as soldiers, we had trained to go to war. To be told we were not required was disappointing. In the following weeks, we could only watch on the news from our parent base in Hamelin, Germany, as British soldiers wearing their nuclear, biological and chemical warfare suits searched for the weapons of mass destruction our government at the time believed were being developed by the Iraqi leader, Saddam Hussein.

After a month, with the Iraq invasion in full swing, out of the blue, the rest of my unit got a call to go to the Democratic Republic of Congo (DRC). Two militias in the DRC, called the Lendu and Hema, were battling for control of a town called Bunia after Ugandan troops had withdrawn following the signing of a peace agreement. This fighting resulted in the Congolese police fleeing. There had been killings and maiming of innocent civilians by militia in the region and aid was required urgently to help the local people.

After the horrific murders of civilians, the United Nations secretary-general Kofi Annan intervened. He called for the establishment and deployment of a temporary multinational force to secure the area and build a new 100 m by 100 m concrete apron pad off the current potholed runway to land and offload more aeroplanes so that vital aid could reach the region faster. We Royal Engineers were the chosen troops, and the name of the operation was Operation Coral.

We flew into the DRC via Uganda in June 2003, naming our small, tented camp Rorke's Drift in tribute to the Victoria Cross recipient, Lt John Chard, Royal Engineers, who fought in the nineteenth-century battle made famous by the film *Zulu* in 1963.[12] French troops were already on the ground, and they had designated us a patch where we would set up our camp. Our accommodation for the next four months would be several canvas tents sleeping eight men, each with military-issue cot beds.

The French had installed some barbed-wire fencing around the perimeter, but we knew it would not be enough to stop any militia from using wire cutters to gain entry if they so desired, so we set about installing our own razor-wire fencing. It came coiled up in 10 m lengths and was placed around the perimeter of our camp at 2 m high and 3 m wide. After nightfall, a two-man roving sentry patrol walked the internal side of the perimeter fence with night vision goggles. Patrols swapped personnel every two hours and ensured that no one attempted to cut the razor wire, which would take them a good hour to do.

Five Congolese police officers had set up a small station in a brick building next to our camp where two prisoners were currently locked up. One morning, we were about to start working on the new apron when we saw the police whipping both men and goading the two of them to fight each other. They looked exhausted and bloodied from the punishment. We briefly turned a blind eye, but the cracks and the screams got louder. A few of us picked up our rifles and walked over to see what was going on. An overweight police officer was sitting down on a chair sweating from his brow and pits. He explained, laughing, that the men were thieves and were being taught a lesson. After some negotiation, he eventually agreed to cease the torture to keep the peace. No sooner had we got back to our position than we heard a commotion behind us. One of the prisoners had broken free and was sprinting towards the razor-wire line. Upon reaching the perimeter, he dived headfirst from a mound of dirt just inside the wire, clearing the height, but not the width. He struggled like crazy, ripping off his torn, bloodied clothing and tearing his flesh through tears and screams as the wire cut into him. After two minutes, and to our surprise, he made it through and took off, near naked and bloodied, across the grasslands. Sadly for him, an hour later, he was dragged back, but I had to admire his efforts. More pertinently, upon seeing this adrenalin-fuelled man attempt to leave our cordon, we knew that our containment fence needed to be improved upon, so it was revisited and addressed to do so.

A week or so later, I awoke to automatic machine gun fire and shouts from the sentries on patrol of, 'Stand to! Stand to!' I got dressed in the complete darkness as did the other seven men in the canvas tent, putting on the first things I could find, which were a pair of shorts, a military shirt, my boots, helmet and body armour. I then looked out of the non-protective tent to see tracer fire from the French machine gun inside our camp fizzing out over our heads and the runway into the distance, well past the protective fence line.

I was told to go with Mojee, a Fijian soldier, to our pre-made trench about 100 m away on the edge of the internal perimeter fence line. He had the general-purpose machine gun (GPMG) and I had the night-vision sight. We ran down to the muddy trench, which was 4 m wide and 1.5 m deep, where Mojee cocked his GPMG upon fixing a firm footing. I checked through the night-vision sight to see dozens of eyes glinting back at me from about 20 m away.

After a split second of panic, I realised stray dogs were sitting on the other side of the razor wire, attracted by the smells of the fire where we threw all our old food and waste from the cooks' tent. We never saw who had fired the first shots towards our camp, but the French were on hand to return fire from their elevated position, which soon drove away any attack. It made a change for the French to come to our aid!

There were no further attacks in the remaining weeks and no one, not even wild dogs, managed to break through the reinforced cordon we'd set out. The concrete 100 m × 100 m apron adjoining the runway was completed in time with a pat on the back from a high-ranking French general, who flew in specially to announce the official opening.

This military anecdote shows how important it was to cordon off and contain our camp to prevent local militia intruders getting in. By installing the wire fence around the site, we were able to ensure the safety of those in the camp as well as prevent any prisoners from escaping.

Pest cordon and contain

When it comes to using barriers to keep pests out, or contain them in some way, it is important that you identify what pest you need to block out before deciding on the methods and materials you require. Not all pests can be cordoned off to prevent their destructive nature. For example, woodworm is difficult to barricade against, especially once it has carried out its ingress into the wood.

It's not only pests that require containing; food and clothing may also need to be quarantined. We are probably all guilty of leaving open plastic-packaged foods, such as rice, cereals, flour, pasta and spices, in the cupboards. Over time, this creates a perfect, quiet, undisturbed nesting haven for insects such as Indian meal moths and other stored food-product beetles, such as weevils, biscuit beetles, grain borer beetles and more. Protect packaged foods such as rice, pastas or dry pet foods by putting them in hardened containers, especially if you are buying them in bulk. Thick plastic containers will stop rodents and insects accessing these foods.

The same goes for clothing. If you are not currently wearing your woolly Christmas jumper or ski wear, get it quarantined in a sealed bag or suitcase. This stops textile-munching pests such as moths and carpet beetles nesting and using the garments as a breeding and feeding ground before migrating throughout the property.

Insects

Encasing mattresses with a mattress cover to prevent bedbugs is popular, but my experience tells me that bedbugs will generally gather around the wooden slats of the bed or on the back of the headboard. Not only that, but they can lie dormant for up to a year without feeding, so it is crucial to have the insect treated and eradicated rather than trying a containment method.

Door and window meshes are useful to protect against flying insects, especially in working kitchens. Flies spread disease, and if a customer falls ill after consuming food served in a restaurant, it is likely to damage the restaurant's reputation.

These cordons are barriers to keep pests out, but when natural fibre-feeding pests have contaminated a garment, a containment method to keep pests *in* will stop them spreading further. Moths are a nuisance and when they start eating your favourite cashmere, you may feel they have gone too far, but it's not the flying moths that cause the damage; it's the larvae. Treat your garments by washing them at over 60°C if possible, ensuring not to shrink delicate items. For the delicate items, freeze them for seventy-two hours, and then quarantine them in suitcases or vacuum-packed bags if you are not going to wear them for a while. This is important to protect them.

Rodents

Mice and rats are fantastic scavengers, and a pantry filled with food delights is a goldmine for these creatures. Mice can squeeze through the thinnest of gaps, so the pantry door is unlikely to stop them. Rats, once given a free rein, will be destructive and may gnaw through thin wooden doors, so watch out for any shavings around the wooden barriers.

Birds

Most bird species in the UK are seasonal, meaning they will only be present for certain times of the year, usually in the warmer months, but some, namely pigeons and seagulls, will be found in and around the same sites all year round. When these birds find a cosy, sheltered nesting site with a food source close by, they tend not to move too far from that location.

Redundant properties or buildings with roof spaces that have easy access are favourite locations for birds, as are heating and ventilation ducts on the roofs of commercial premises. I have seen over sixty pigeons gain entrance to a property's roof space while elderly people were living in the home below. The loft was in chaos and stank with the mess these birds produced. The problem continued for years, until eventually the weight of the pigeon guano and the small leaks in the roof took their toll. The ceiling plasterboard caved in, resulting in the ceiling and the pigeons dropping down into the living space of the property. Fortunately, the residents were not upstairs at the time of the ceiling collapsing.

It cost this client over £2,000 to fully remove the pigeons and carry out a biohazard clean alone, not to mention the structural costs. My advice is to check your loft space monthly for any pest activity and block off all entry points in the roof. You may also find insects or rodents nesting. If it's wasps you find, please call in a professional. If you find bats, these are protected and, again, must be left alone.

Foxes

You can create a barrier against foxes around the perimeter of your garden, but this will come at a cost, and I do not just mean financial. Aesthetically speaking, caged fences make a garden look unsightly, more like a prison yard than somewhere to relax. They may also lead to complaints from your neighbours. This is something to be aware of if you're considering fox-proofing your garden areas.

Things to note

- When you're using barriers to keep pests out, think about the best material to suit the job. If the pests are rodents, steel mesh is best as their teeth are unable to penetrate it. Concrete is another material that rodents will struggle with. Wood may work for mice and possibly rats, but if the wood is soft and thinner than 3 inches, then larger rodents such as squirrels and rats will likely regain entry quickly, creating new holes.
- Quarantine clothing in sealed vacuum-packed bags, or at least bin bags tied at the top so that no textile-feeding pests can get to them.

CASE STUDY: AN OFFICE INVASION

I was in an estate agent's office in London, which was plagued by a stored product pest called a saw-toothed grain beetle. They were everywhere, crawling out from gaps around spotlights and incoming water pipe services leading from the ceiling, but I did not find any stored foods in the office space that would create this size of infestation.

There were flats above the office, so I went upstairs, knocking on doors. One flat had been unoccupied for over a year. Immediately, I knew this was the epicentre of the infestation and I needed to gain access to this flat to stop the problem reoccurring.

The estate agent was able to access the upstairs flat and we found the kitchen was covered with crawling beetles, feeding on opened bags of flour and cereals. These foods were also acting as a breeding ground for the pests.

It cost the estate agent company over £400 in pest-control fees, not to mention the time and effort to remove the loose foodstuffs. This problem could have been resolved easily if only the foodstuffs had been contained.

Summary

The military story at the beginning of this chapter highlights the need not only to keep intruders out, but also to keep things (or people) in. In a similar way, cordon and containment is necessary to prevent a pest infestation growing to uncontrollable levels, so we have looked at key ways to do this depending on the pest infesting your home or workspace.

FIVE

Control Of Mammals And Birds

Gaining control of a pest infestation quickly is crucial to prevent the situation from escalating and spreading. If things are left to fester, a small infestation can rapidly migrate throughout your building, and then into neighbouring properties, affecting people's lives and relationships, especially if the finger of blame is being pointed at you.

Control, the fourth of the 6Cs, is a vast subject, so I will cover it over the course of two chapters. In this chapter, we will focus on how to control an infestation of different types of mammals and birds, while in Chapter Six, we will deal with the control of insect infestations. I will recommend products you may want to use in each situation, but please be aware of health and safety issues and always read the instructions before using these or any other treatment products.

Let's start with an example of the importance of control in a military context.

Military control

In 2005, after finishing my advanced electrician's course, I agreed to a six-month posting to the British Army Training Unit Kenya (BATUK). My duties were to test the electrics in all of the accommodation blocks and repair any faults that might occur during my time there.

Most of the work was in a place called the Nanyuki Show Ground, which was a holding camp the British troops stayed at before they carried out hot-weather exercises in the Archers Post training area, sometimes referred to as Archers Roast due to the extreme heat. Behind the camp was a fantastic view of Mount Kenya. After three months of staring at this beautiful mountain, myself and Chris, an infantryman who worked on camp, decided to climb Mount Kenya's third highest peak, Lenana. This peak was the highest one accessible to us as the others required pitch climbs with ropes and other equipment we were not experienced in using.

After five hours of hiking uphill with a local guide, we came across a young woman being violently ill. Her guide was struggling to carry her down as she was clearly suffering from the onset of altitude sickness. We knew we needed to help.

A British military exercise was taking place up at Archers Post where there was a military Lynx helicopter on standby for emergencies, as well as to provide simulation training. As part of our preparation for the climb, I had called the British Army pilots the day before we set off to ask them for assistance should we get into any trouble, which they happily agreed to. Chris and I had taken a smoke grenade each as a location marker for the pilots if required. As a last resort, we would call the pilots to assist us with the sick woman, but when we assessed her, we diagnosed her condition as altitude sickness and realised it was not a life-and-death situation. If we could carry her back down, she should recover naturally.

We began the journey down the mountain to control her symptoms, carrying her piggy back and stopping occasionally as she was trying to be sick or in tears of pain. It was amazing to see how quickly the young woman recovered from being so ill and disorientated when we had dropped a mere 30 m in altitude. We left her with her guide as soon as she was able to walk normally and seemed compos mentis, then continued back up the way we'd come and spent the night at the base camp.

At 3.30am, we woke to head for sunrise at the summit, witnessing it with Kilimanjaro in the distance. My companion had complained of a headache on the final push, another symptom of altitude sickness, but we made it back down safe and sound that day and his symptoms, like the young woman's, were rapidly relieved by the descent.

It was vital that my companion and I took precautions before what could have been a dangerous climb. We took control of our safety by notifying British military pilots of our plans and our route before we set out and arranging support from them should we need it, but when we met the guide carrying the sick woman, we knew we had to control her environment too if she was to recover quickly. This led us to think logically, change our plans as necessary and help get her down to a lower altitude.

How does the principle of control apply to the pest environment? In this chapter, we will analyse how to control mammal and bird pests in your home or workplace. This is by no means an extensive analysis, but it will give you an idea of the actions you could take to eradicate the infestation you are dealing with and prevent it from escalating further.

Rodent control

Eradicating an infestation is the main objective in pest control, so you can get on with your life without one more unwanted stressor added to the pot, but the control methods you require will depend on what pest you are dealing with. You can solve pretty much every pest infestation by yourself; the key is to find the source of the problem and know how to rectify it.

Let's start by looking at arguably the most common animal pest: the rodents. The Latin term for gnaw is *rodere* and this is where the word 'rodent' originates from. Their incisors (the two top and bottom front teeth) never stop growing, so they gnaw to file these teeth down throughout their life.

Mice control

The two most common species of mice in the UK that are seen as pests because they invade our personal space are house mice and field mice. House mice are grey or dark grey in colour and will predominantly be found inside, but in the warmer months in the UK, they may dwell in sheds and garages, especially if there is an easy food source close by. Field mice are more yellow and gold in colour with a distinctive white underbelly.

House mice are the smallest rodents my company treats. They are generally 7–10 cm long with a tail of similar length to the body. Their nose is pointed and their eyes and ears are large in comparison with the head. Their feet are relatively small, distinguishing them from a baby brown rat that has big feet. Their droppings are a fine spindle shape, so you could compare them to grains of black rice.

Mice can burrow but prefer to use cubby holes that are already present. They will scale up and down electrical cables and water pipework 20 mm in diameter or smaller to move through a building, and may nest in insulated walls and roof spaces. Most people with mice in their properties will complain of soft scratching noises in the cavities, especially during the night when they're trying to sleep.

Mice are sporadic feeders and have evolved to feed off crumbs on the floor. They know that breads and chocolate are a safe food source for them in the UK, so the best rodenticide baits are ingrained with wheat or chocolate scents. This is my opinion, but others may disagree.

Treatment

- Survey your property to establish possible entry points and nesting locations. Always check the incoming boiler pipework, water pipes for all sinks where they enter the living space, under the stairs, and gas and fuse board wires where mice can enter the living space. Holes cut in walls to allow these services to enter a building are usually the most common entry points. Block these holes with steel mesh, wire wool or a hardened material such as cement wherever possible, but you may need the help of an expert. It is essential, though, that you eradicate the infestation before blocking any entry or exit points.

- Be mindful of external gaps through which you can fit a biro pen lid, and any other gaps under doors as well as air vents under steps or on a side wall, as these may have holes big enough to accommodate a mouse. This is a common entry route for mice from the outside, so again, holes will need to be dealt with once the infestation has been eradicated.

- Locate droppings. If there is a large concentration of droppings in one area, it is likely this is near an entry and exit point. Mice are incontinent and can produce up to eighty droppings a day, so large numbers in one area tell you that it's a commonly travelled path. This indicates where you will need to place your rodenticide baits and/or traps.

- Look for smear marks in the areas above and behind kickboards (the plinths at the bottom of kitchen units). These indicate where the mice constantly rub their fur when jumping over objects, leaving a dark residue. Again, this is where you want to concentrate your rodenticide baits and/or traps.

- Establish why the mice are present. Is it because of pet food being left out or young children dropping food? Make sure to contain all foodstuffs and keep your floors clean. Can the mice climb on to the kitchen worktops and get into the toaster? If so, make sure to move anything they can scale such as fridge electrical leads.

- Consider whether they could be coming from a neighbouring flat or house, entering to mark their territory as your property covers their territorial patch.
- Use appropriate traps, rodenticides and bait stations that are approved to national standards. Always read the safety information for the poisons you will be installing.
- Install the rodenticide baits appropriate for the situation in safe, contained locations out of easy reach of people and pets. Common areas include behind the kitchen kickboards, in the loft and/ or cellar, behind bath panels or any other likely location that the mice move through.
- If the mice are not eating the rodenticide you have put down, they are likely feeding from another food source nearby. Be mindful if there is a pet in the property to remove the bowl when the pet is not feeding. If the neighbours have a pet, advise them to do the same.
- I advise using a minimum of two rodenticides, one in a soft palatable form and the other a hardened granular form. Don't use solid wax blocks. I have found that the mice I have treated in London tend not to feed from these, which may be due to the difficulty of consuming them and the time it takes. The active ingredient (poison) I favour at the time of writing is Brodifacoum.

Rat control

The most common species of rat in the UK is the brown rat. Brown rats live in large, hierarchical groups, either in burrows or in subsurface places such as the sewers. When food is in short supply, the rats lower in the social order are the first to die. A rat's maximum life span is three years, although most will only survive for one year, usually due to predation, disease, environmental conditions, interspecies conflict and rodenticide control. If a large portion of a rat population is exterminated, the remaining rats will increase their reproductive rate and quickly restore the old population level.

Rats have an acute sense of hearing, frequently using ultrasound to communicate, and are particularly sensitive to any sudden noise. A lot of rat infestations access properties by travelling up the waste pipe, which runs from the back of your toilet to the sewers, creating or finding weaknesses in that pipe to gain entry if the property has cavity floors. So if you have a toilet in the middle of your property you know you have a sewer line leading to it in the cavity of your building. They may also burrow and nest under garden decking or sheds. Female rats tend to do this to prevent their young being killed by adult rats in the sewers.

Treatment

- Survey the whole property, including the perimeter of the building for rat burrows leading to the foundations. Check for open-ended drainpipes that rats may climb to access the roof. Flattened grass that leads under decking and/or smooth soil where the animals' underbellies are creating pathways are tell-tale signs of the presence of rats, showing you where to concentrate your chosen method of treatment.

- Check for gaps in air vents under steps or in external side walls as these may have holes big enough to accommodate a rat. Block these with steel mesh or concrete, if possible, but only after the infestation has been eradicated.
- Scan for manhole drain covers at the front or rear of the property. You may require a pest-control expert to access these if you suspect or they establish a break in the sewer line.
- Find out why the rats are present. Food sources such as bins, bird feed, composts and fruit trees will attract rats, as will ideal nesting locations like decking or a drainage line under the property. If the property has an understairs toilet or one not connected to an external wall, there will be a sewer line running below the building's cavity floor or wall. Sewers are how rats predominantly travel between locations, so the sewer line could be damaged under the property or at the back of the toilet allowing countless rats to enter. Make sure any redundant sewer lines are capped off properly by an expert.

- Members of the public are not permitted to bait outside burrows with rodenticides, and you should never leave rodenticides outdoors on show or accessible to natural wildlife. Birds or foxes are especially likely to dig up bait, so please inspect baits regularly and do not place rodenticide if you think there is any chance that wildlife can get to the poison.
- The only approved method for baiting outdoors is for the rodenticide to be secured in a hardened, tamper-resistant box that only rodents can access.
- Open rodenticide bait trays are only to be used in contained areas inside the property, such as behind kitchen kickboards, in cavities, in loft spaces and other areas you believe to be isolated from grabbing young hands and curious pets.
- Rats may have neophobia, which is a fear of new things.[13] This will make them nervous and cautious of anything being constantly moved, so leave the rodenticides in place for a good week or so to ensure the animal is comfortable with their surroundings.
- Identify and block entry points. Common places are under the stairs, behind the kitchen kickboards and burrows from outside, so be mindful of decking. Look for sewer lines running below the property and non-capped-off vertical rainwater gutters that the rats may climb.

Remember the old saying 'As quick as a rat up a drainpipe'.

- Rats will feed on most things. I would recommend chocolate spread as a lure on rat traps, and both soft and hard rodenticides are good, though I prefer soft palatable baits. At the time of writing, the active ingredient (poison) I recommend is Brodifacoum. How much the animal eats will dictate how long it will take for it to be eradicated. It can take up to a week and may be longer.

Grey squirrel control

Two squirrel species live in the UK – the red and the grey. The red squirrel is native to the UK, whereas the grey was brought to Europe in the late nineteenth century. The red squirrel is protected under the Wildlife and Countryside Act 1981, whereas the grey squirrel is not. This is due to the grey not being native to the UK and the fact it carries the pox virus, which is fatal to the reds. The disease can be spread by both species feeding from the same bird feeders. This is why red squirrels are no longer found in most parts of the UK. The virus cannot be spread to humans.

Despite its cute appearance, the grey squirrel is a pest that can cause much damage and give you an early wake-up call if it nests in your loft space or chimney. Grey squirrels can also become aggressive if their territory is threatened, especially if they have young in the nest.

Squirrels will nest in loft spaces that have a high clearance; it's very unlikely they will nest between floors due to tight access. If you're hearing noises between your floorboards, it will likely be rats or mice.

Treatment

- Inspect your property for entry points. Look for gnawed holes in the soft wooden fascias up high. Focus on the loft and roof areas, around soffits and cladding, and block holes where possible with steel mesh or new cladding. Replace slipped roof tiles. Dirt smear marks on the wall or guttering are also evidence of squirrels' travel routes.
- Do not block access holes until you are confident the infestation is removed. Squirrels will create new holes to regain entry, especially if they have young inside. It is illegal to trap animals inside knowingly as it causes them huge amounts of stress, and they can then do untold damage to your property.
- Bird feeders and other accessible food sources will draw the grey squirrel to nest close by, so make sure to contain foodstuffs and only fill bird feeders when you can monitor the creatures using them.

- Trees near your property may require cutting back to prevent squirrels accessing the roof of your dwelling.
- Where possible and when access allows, install powerful break-back traps called MK6 Fenn traps where the squirrels nest inside your property. Do not use these traps outside. Please read the instructions carefully as these traps can cause you harm when you're setting them if you carry out the work incorrectly.
- Outside traps called tube traps can be fixed to the guttering or to squirrel runs, which are their paths of travel. Please note, every trap must be fixed down to prevent a squirrel running away with it if caught by its tail or another body part. This is rare, but it can be distressing for the animal and you.
- Automatic traps fixed in position outside your property work well if you suffer from multiple squirrel infestations throughout the year. Ensure that all traps are well away from children's grabbing hands and curious pets.
- No pesticides are allowed for use against squirrels, who spend a lot of their time outside and may die on open land from the poisoning. As birds feed on the deceased animal, they may receive secondary poisoning. We need to protect our wildlife.
- *Do not* install non-tubed open traps such as MK6 Fenn traps or place pesticides outside. This is

illegal and can kill non-target species. The last thing we want is to kill a cat, small dog or our natural wildlife.

Fox control

Foxes are natural scavengers and are an ever-increasing sight in major towns and cities. Most urban foxes see humans as a food provider.

Foxes are protected under a series of wildlife laws. At the time of writing, anyone harming a fox is liable to six months imprisonment and/or a £5,000 fine per animal.[14] The fox is sometimes referred to as vermin, but it has never been categorised as such by the Department for Environment, Food and Rural Affairs (DEFRA).

However, foxes are *not* protected against shooting. A single shot that culls the animal instantly is the only legal form of eradication. Some pest controllers may charge large fees to cage-trap nuisance foxes, which are then set loose miles away, but please don't take this option. A fox dropped off in a strange territory (known as hard release) will find itself in competition for food with resident foxes. This puts the animal under undue stress and is therefore almost certainly an offence of cruelty under the Animal Welfare Act 2006 and is condemned by DEFRA.[15]

Treatment

- Make your garden fox-proof. Install fencing and steel barriers around the garden's edge to stop new foxes entering, but be aware that this will be costly and possibly not pleasing to the eye.
- Remove any shelter, food sources or general clutter that may attract foxes. Overgrown gardens which provide adequate shelter, decking and sheds placed on soft soil which the fox can dig below and create a den are all favoured spots for the animal. Use wheelie bins for food waste to prevent mice and rats, which the fox will see as food sources so will stay local to the area.
- It may be the neighbour's property that is causing the animal to stay local, so work together with your neighbours as a community. This will be advantageous in many ways in the long term.
- Ensure any hutched animals such as rabbits, guinea pigs and chickens are kept in secure bordered areas to prevent foxes gaining entry.
- Use a deterrent scent like Scoot. Some of my customers even put their urine in jars around the garden to deter the fox.
- If the foxes are still present after you've carried out all these treatments, the final option is culling using a professional fox specialist.

Pigeon/seagull control

Pigeons and other wild birds are protected under law. The Wildlife and Countryside Act 1981 declares it is illegal to kill or injure any wild bird, including pigeons and seagulls, unless general licensing regulations are complied with.[16]

Treatment

- Identify the attractive features in the local area that are enticing the birds to stay close. These include bread or other food left out for the birds, bird feeders, ornate water features and safe shelter. If they're found, look at ways to remove these attractions or prevent the pigeon or seagull's access to them.
- Remove or contain food sources such as seeds or fruits as pigeons are herbivores. Food is more important to them than the threat of predators.
- Never feed pigeons or seagulls, otherwise they will keep coming back.
- Install bird spikes, nets or wire to prevent pigeons resting and nesting around your property.
- Pay particular attention to open gaps around the roof of your property that may allow pigeons to nest in the loft space. You may need to work with an expert to block these gaps, but make sure no birds or eggs are trapped inside.

- You can install a figure of a life-size hawk in your garden as pigeons are fearful of larger birds, but through experience, I know that birds quickly get used to ornate objects.

CASE STUDY: PIOUS PIGEONS

I was called to a church where two pigeons had begun nesting. They were causing a racket and occasionally fouling on an unlucky bride and groom who were exchanging their vows down below.

I carried out a risk assessment to make sure that any errant air rifle shots would not damage the listed building and windows – not that I missed that often, thankfully. When I was ready, one of the pigeons flew to the large cross at the front of the building. The vicar was standing behind me as I set the crosshairs of the rifle's scope on the plump chest of the bird, but as I squeezed the trigger, I suddenly realised what I was about to do. I took the pressure off the trigger with the rifle still firmly fixed on the bird.

'If I kill this bird on the top of the Lord's cross in His own home, will I be struck down?' I asked the vicar.

'I don't think so, Mark, but I may take a couple more paces back, just in case,' he replied.

Bang! A puff of feathers and the bird fell to the floor. Fortunately, there was no bolt of lightning scorching me to a cinder, but I did feel pretty shitty. Pigeons are monogamous birds, so its female partner was now a widow. Not for long, though, as I shot the female bird minutes later.

For the record, I don't like killing animals. The only time I personally think it is OK is if I'm planning on eating the animal afterwards. I could have rustled up a pigeon pie with the two birds, but I think it would be a lot of work for the limited amount of meat on them.

Summary

The military story at the beginning of this chapter highlighted how to control a situation by knowing what to do and being prepared for the worst-case scenario. The same approach works well when we're dealing with an infestation of rodents, foxes or birds.

Rodents are the most common pests in major cities. Victorian homes with their abundance of cavities make them a great place for mice and rats to nest, so control is key to stopping a small infestation becoming unmanageable. Once the infestation of any pest is under control, we then need to eradicate the pest's source of entry and whatever was attracting it to prevent the infestation from happening again.

Christmas Day at the Baron hotel in Kabul, Afghanistan, with Steve (I'm Santa); the local Afghan people always saw him as an elder and gave him enormous respect, especially after he helped out Tony and me when we broke down in the middle of Kabul

Operation Coral 2003 in DRC with 42 FD Sqn. From left to right: Johno, Mickey, Macca, Mojee, Ratsy and myself

Operation Coral, DRC: fixing the fence line after a local prisoner escaped, 2003

Heading up Mount Kenya, just before we came across a woman suffering with altitude sickness, 2006

The sunrise view from the summit of Mount Kenya, Lenana peak, 2006

Skydiving with the British Army, 2007

Working as contractor in the Middle East: here I am at a Forward Operating Base in Afghanistan, having a well-deserved cigar with Tony after this tower construction job was completed, 2011

A favourite pastime of mine, surrounded by mountains, Nepal, 2013

On the summit of Mont Blanc with the PestGone banner: climbing mountains gets me away from the London rat race and always helps me reassess the business, 2020

A wasp nest treatment in a church in Peckham, southeast London, 2018

At the SME National Business Awards ceremony at Wembley Stadium, 2021

Speaking at the Churchill War Rooms, Whitehall, about my journey into pest control, 2021

A sight we see occasionally after the summer: a severe bedbug infestation on the Old Kent Road, southeast London, 2022

Signing a new pest-control contract for a multi-national corporate client, 2022

SIX
Insect Control

In this chapter, we will continue our journey into the treatment processes used to control pest infestations, focusing on insects. Although they are not actually insects, but arachnids, I have included spiders in this chapter.

Keeping a cool head

After leaving the Army, I worked as a security contractor in the Middle East. I had been tasked with a job in Baghdad, Iraq, with another colleague, Pete, working on an American military base in the capital. Iraq seemed to have quietened down on the news and the threat levels didn't sound as bad as in Afghanistan. Even so, the Turkish construction company that we were working with on the project had informed us that a professional security team would be collecting us with weapons and body armour in steel armoured vehicles for our safety. We landed at Baghdad International Airport and were met at the arrivals gate by a local man holding up a sign with our names spelled – almost correctly – as 'Petr' and 'Merk', with our company name, which was spelled correctly. Our driver walked us to a deserted underground concrete sand-brushed carpark. His was the only car parked there in the middle. The windscreen was smashed, and the paintwork rusted away on many parts of the beaten-up vehicle. I questioned the driver, who was a local Iraqi man, and he pointed out the names of the Turkish company and our business, which were correct. Pete and I assumed the Turkish company had saved money on an armoured escort and sent this guy for the pick-up instead. We tried calling the company and our colleagues, but we couldn't get through to them. We asked how far it was to the drop-off, and it was only a 20-minute drive with no stops, so we agreed to the move, especially as no one would suspect the two of us travelling in this clapped-out old car, making us more discreet and helping us not to stand out.

We left the airport and began driving through Baghdad. It was classed as the green safe zone, under local police and military control and seemed quiet. Then, 10 minutes into the journey, the driver pulled up outside a mosque. He left the vehicle with a slight jog towards the back of the vehicle and into the building, leaving me and Pete shouting after him. Within 30 seconds the mosque's doors had opened and the car was surrounded by dozens of men. We thought we had been set up but tried to remain calm.

The men were looking into the car, trying the door handles and banging on the windows. Looking back, I think this was mainly because it was blocking the main exit to the mosque. I had locked all the doors when the driver left and held one of my bags up against the windows to try and hide my Western complexion in case any of these people were hostile. Our driver had left the keys in the ignition, so I decided to just drive off in his car if it would start. As I jumped from the back into the front seat, the driver returned clutching a pack of cigarettes – which, it turned out, is what he'd run off to pick up – and banged on the window to be let in. I wasn't sure whether to open the door or not, but I did move over to the passenger side, and he jumped in before I locked his door again, shouting at him to, 'Fucking drive!' As we sped off with plumes of black smoke bellowing behind us, Pete and I gave the driver an earful.

We arrived outside the American base, where we changed vehicles to a 4×4 white Toyota Land Cruiser, driven by one of the Turkish construction company's staff. Still a little upset, we began giving him a hard time as we passed the security on the base. We arrived on camp and were met by the manager of the Turkish firm who apologised for our collection and explained that the armoured vehicle road move with a close protection team had been cancelled that morning, so the local driver was the best they could do. I was dubious about the cancellation, suspecting it had more to do with cutting costs.

After checking my pants were not soiled, I could look back on the incident and see that remaining calm, locking the doors and staying in control of the situation, while realising we had the keys to help remove us from any danger, were the right steps to take. We could have opened the doors and confronted the local men banging on the windows, who were likely only annoyed that our vehicle was in the way of the exit for the mosque, and this could have made things worse. This experience also taught me not to deviate from the original plan if there is a chance that my or my colleague's safety could be put at risk, and that staying in control is key – in military or pest-control contexts.

Common UK insect pests

In this section, we will look at the key insect infestations you may encounter in your home or business. These insects include ants, bedbugs, carpet beetles, cockroaches, fleas, flies, gnats, moths, wasps and woodworm. As always, remember to apply health and safety precautions when you're handling the treatment products I recommend. Follow the instructions for any product you buy over the counter or online and ensure you use it safely so no harm comes to the people and/or pets or aquatic life you may have in the property.

Ant control

The common black garden ant is fast moving and will run away from human confrontation. Flying drone ants come from the same nest as the crawling worker ants. The queen specifically nurtures certain female ants to continue their species. Potential new queens are more valued by the colony than drones, who die after mating.

The black ant will enter properties for food. They eat leftovers, sticky drink spills, jams and soft fruits. They also eat honeydew from aphids that the ants farm around gardens and vegetation as well as feed off other insects.

Black garden ant colonies can have up to 10,000 workers. The queen ant is around 9 mm long with dark brown/black/reddish legs and antenna, and has been known to live up to fifteen years in ideal conditions. This means that every year, the ant problem will return if you don't treat it early. There is one queen ant per colony for the garden ant species. Once she is eradicated, the colony cannot survive without her, and so it collapses. Any potential queens and drones in the nest still being nurtured by the workers may make it to adulthood and then create their own nests.

Treatment

- Before embarking on a course of treatment, ensure the ants you have are black garden ants. Other ant species may need to be treated differently due to having a different hierarchy system.
- Survey the whole property, inside and out, to identify the level of infestation and nest location. Particularly, check regularly watered plants and drain covers.
- Ensure any sticky fluids or sweet-tasting foods are sealed.
- Don't use a kill-on-contact insecticide such as permethrin because eradicating the worker ants will not eliminate the source, which is the queen who is in the nest, producing more ants. Instead, use a sweet-tasting insecticide with a slow-working active ingredient (toxin/poison/insecticide) such as fipronil. Place this inside a protective container or pot that only ants can gain access to. It only takes a low number of ants to distribute the poison to the nest in a process called trophallaxis, as the queen and larvae feed from the regurgitations of the workers until the entire nest is killed off. It may take up to four weeks, possibly longer, to fully eradicate the nest.
- If you are plagued by pharaoh ants, do not use a kill-on-contact insecticide or cleaning detergent to eradicate them. Pharaoh ants will nominate

other ants in the colony to be queens in a process called budding when they are under attack. These new queens will then start reproducing and laying eggs, which increases the number of insects significantly. Use a food insecticide such as biopren instead. For pharaoh ants, check the heating system rooms where boilers or water tanks are located in a property as pharaoh ants will likely be dwelling around these locations.

- You may have a problem with brown ants who, like black garden ants, only have one queen. For these ants, use a sweet insecticide gel treatment, but if this is not successful, you will need to get access to the cavity of the building that is harbouring this pest and get it fogged. If living in a block of flats, I would advise speaking to the neighbours to find out if they are dealing with this pest too. The fogging process will only be successful if all properties in your building are treated.

Bedbug control

Adult bedbugs resemble a small brown disc, measuring up to 6 mm in diameter. A bedbug is a wingless insect, but it can crawl up most vertical surfaces. It resides in cracks and crevices close to a host, which, in nearly all cases, is a human, so it's imperative that you check headboards, wooden drawers, wooden bed slats, mattress seams, loose wallpaper and skirting boards adjacent to the bed. Bedbugs are usually brought into the property on clothing, bags and luggage, or on infested second-hand furniture and beds.

Bedbugs are mainly active at night. Like fleas, they are parasites that feed on blood in localised areas on the human body, resulting in multiple bites in one location.

Treatment

- Check the bed frame, headboard, bedding and mattress for adult bedbugs. Check for blood spats on the bedding and/or brown/black dots, which are evidence of this pest's secretion. If you find these signs, strip the bed and wash the bedding at over 60°C.
- Put delicate items located next to the bed that may be infested in a freezer for seventy-two hours. The cold temperatures will eradicate this pest and its eggs.
- Pull the bed out and vacuum around the headboard, including its back, especially where folds of materials meet or are stapled. Immediately empty the vacuum cleaner bag outside after doing so.
- A steamer will kill off bedbugs.
- Off-the-shelf insecticides can be purchased, but my customers report that they do not have much luck with these against this persistent pest.

- Professionals can use specialist heat equipment, but this is costly due to the amount of time they're required to be on site. This treatment in my opinion needs to be combined with an insecticide to ensure complete eradication.
- Insecticides that leave a long-lasting residue behind work well as a treatment as they also kill off nymphs (newly hatched bedbugs).
- Fogging is not very successful at treating this pest, so do not try flea bombs or other such aerosols in the hope that they will solve the problem. This is because bedbugs hide in cracks and crevices where the fogging insecticides may not access. I have had customers tell me they sprayed a fogging insecticide from close range directly on to a bedbug, but it didn't kill the insect.

Carpet beetle control

The carpet beetle is 2–4 mm in length and looks like a small, mottled brown, grey and cream ladybird. The larvae, known as woolly bears, are covered in brown hairs and roll up if disturbed. As they grow, they moult, which may be the first sign of an infestation.

Carpet beetle larvae hide in dark, undisturbed areas and feed on organic material, feathers, fur, hair, silk or wool. The adults tend to wander along the pipes from roofs into airing cupboards, which house the clothes and blankets that constitute their food and nests to lay their eggs in. The damage caused by carpet beetles consists of well-defined round holes along the seams of fabric where the grubs bite through the thread. Carpet beetles can fly from room to room, causing the spread of an infestation.

Treatment

- Check natural-fibre clothing, particularly wool and silk, that has not been worn or rotated through the wardrobe in a while. This is a food source for carpet beetles. Treat these clothing items with heat by washing on a hot wash if possible, or by freezing to kill the beetles off. If the clothing item is not valuable to you, throw it away.
- Check the loft space for redundant birds' nests, cut-offs from carpets made of wool and any clothing thrown in there for storage. Remove and dispose of these items.
- Check carpets and natural-fibre items below furniture and in wardrobes. Vacuum and, if you can, steam carpets regularly.
- You can purchase off-the-shelf insecticides to treat this pest. Treat all areas where woolly bears can be found.

Cockroach control

Cockroaches will feed on anything, including refuse, faecal matter and even their own dead. They can survive for months without food, but they will not live for more than a few days without water and are generally found in inaccessible places close to water and food sources.

We have three cockroach species in the UK – the German cockroach, the Oriental cockroach and the more uncommon American cockroach. A German cockroach is a brownish yellow colour with two tanned strips along its thorax. This pest will usually have one single nest where it deposits its ootheca (egg sac). This makes this cockroach easier to treat than the Orientals, which deposit their ootheca in a number of different locations as they travel.

The German cockroach cannot tolerate cold climates and as a result will not move around outdoors in the UK. These cockroaches will move between connected properties when their numbers are high, which can result in re-infestation in a flat within a housing block.

German cockroaches nest in warm places. Behind fridge freezers is ideal as there is a warm motor and a condensation drip tray they can drink from. Oriental cockroaches are usually found in dank basements and around drains.

Treatment

- Treat cockroaches by using an insecticide gel with a slow-working active ingredient (toxin/poison). The gel can be applied to inconspicuous areas, inaccessible to humans and animals, thus preventing unintended exposure. Install the gel behind kitchen kickboards, along pipework, around the back of fridge freezers and washing machines, and inside cupboards and their hinges. It is eaten by the cockroaches, who pass on the poison to other cockroaches through vomit or excrement. Cockroach cannibalism provides a secondary poisoning effect.
- Where necessary, you can include a secondary method and use an off-the-shelf long-lasting residual insecticide spray in conjunction with the gel treatment, spraying the wall/floor junctions and skirting boards, and cupboards once they're cleared of foods and plates. This will work well for all cockroach infestations.
- You can use fogging insecticides, but only if the property is free of people and pets.

Flea control

The most common flea in the UK is the cat flea, known for biting humans. Adult fleas are small (average 2 mm in length) wingless insects, flat and red brown in colour, with backward facing legs designed for jumping.

Fleas feed on warm-blooded animals, usually pets. The females tend to lay eggs after feeding on the infested animal. The eggs then drop from the pet on to the floor or the animal's bedding. This pest is mainly active where pets sleep and in rooms that are used most often. They live on the pets themselves, in carpets, pet bedding and upholstered furniture.

In a flea infestation, only about 5% is made up of jumping adult fleas. Around 10% of the pests are pupae, 35% are at the larvae stage and 50% are eggs. Consider this when choosing your treatment.

If you don't have a pet, fleas could nest under beds and sofas or where people spend a lot of time. The fleas can then jump on to the person and feed from them.

Treatment

- Ensure your pet has regular flea treatments to control any infestations.
- If you do not have a pet, check to see if your neighbours do and ask them to make sure their animals are treated for fleas. Also check for nearby fox dens. To deal with fox infestations, please refer to Chapter Five.
- Mow your lawn and rake up any grass and leaves because fleas can live externally in warmer weather. Faux turf is also a common nesting spot for fleas, especially if you have urban foxes

roaming locally. An organic insecticide can be used for areas outside a property, so if fleas are known to be in the faux lawn ensure the correct insecticide is used when treating so to not harm any non-target species.

- Fogging works well, but remember that the fogging insecticide will kill adult fleas. It will not penetrate the eggs, which make up around 50% of the infestation, so use residual sprays or powder insecticides in conjunction with any fogging insecticide.
- Wash pet bedding, blankets and other items regularly on the hottest temperature possible. Vacuum frequently, especially carpeted areas and furniture your pet uses.

Fly control

Flies are a family of two-winged polluters found in our homes and workplaces. All flies feed by vomiting saliva on to a food's surface and sucking up the resulting liquid. While doing so, the fly contaminates the food with bacteria from its gut and feet, so it may transmit food poisoning and dysentery, or even typhoid or cholera in countries where these are prevalent. The eggs of parasitic worms may also be carried by flies.

Blow flies known as bluebottles (flesh feeding flies) can lay up to 600 eggs. These hatch in under forty-eight hours and produce maggots (larvae) that burrow into meat or carrion to feed, before emerging as adult flies ten days later. Bluebottles, like other flies, are often found around refuse tips, rotting animal carcasses, dirt and dustbins. They commute from filth to food and carry bacteria on their legs, feet and body. Adult houseflies live for two to four weeks but can hibernate during the winter.

Cluster flies are dark greyish in colour and about 8 mm long with yellowish hairs on their back and overlapping wings. In autumn, they congregate in large numbers in the upper rooms or roof space of a property to hibernate, so they are common in rural areas. My customers sometimes confuse cluster flies with blow flies, thinking they have a dead animal decaying close by their property, but these flies are more sluggish in flight.

Treatment

- Ensure scrupulous hygiene and prompt disposal of all refuse. All foods should be kept covered and dustbins need tight-fitting lids and to be sited away from doors and windows. Windows may be fitted with fly screens.
- Keep meats and other foods covered. Use fly killer aerosols and sticky flypapers to kill flies quickly and an insecticidal dustbin powder for your bins.
- Ultra-violet electric fly killers are considered best practice for all food-preparation businesses.

- Survey your property, targeting cavities, lofts and basements where dead animals may be decomposing, and remove them immediately. Also check bins for rotting organic matter.
- A fogging insecticide applied to internal areas will eradicate adult flies, but not other life stages as it does not penetrate the eggs.
- Drain flies will be attracted to a build-up of stagnant water and/or material close to a leaking tap or drain. This water residue must be dried out. If the water remains, the breeding ground remains, and the infestation will continue.
- For fruit flies, removing the source will remove the infestation. Throw away fermenting fruit. Clean up spills of soft and alcoholic drinks. Deep clean the area where the flies have been seen.

Gnat control

A gnat is a small mosquito common in gardens in the UK on warm evenings. The true window gnat is a slow-flying insect about 8 mm long. Its wings are strongly veined with dark tips and more rounded than those of the mosquito. Gnats breed in sewage filters and lay eggs on rotting food, or they may contaminate homemade wines or honeycombs.

Fungus gnats will always appear more prevalent after a good watering of organic greenery.

Treatment

- Don't overwater your plants. Waterlogged plants will be a prime breeding ground for this pest.
- Fungus gnats are usually attracted to light and will jump on your plants. These can be destructive as they will feed on the root of the vegetation, so turn up soil carefully near the base of the plant to look for the glossy, clear larvae. Over-the-counter or online insecticides containing the active ingredient pyrethrin are good for eradicating the adults and larvae. A light mist of insecticide will suffice.
- Use yellow sticky traps placed horizontally on the soil surface to capture large numbers of the egg-laying adults. Gnats are attracted to yellow and are easily removed on the trap before they can lay more eggs.
- Homemade remedies such as a combination of peppermint, cinnamon and sesame oils in a non-toxic spray are believed to get rid of gnats and other insects that gather around windows.

Moth control

Most moths in the UK are harmless, but a few are seen as pests due to the damage their larvae cause to materials, textiles and stored products. Moths pose no health risks to humans, but they can severely damage natural fibres in carpets, clothes, fabrics, fur and even leather.

What moths are considered pests? For simplicity I have put them into two categories: textile moths and stored food product moths.

Textile moths include the common clothes moth, case-bearing clothes moth and white-shouldered moth. Clothes moths will feed on most natural fibres such as wools, cottons and silks, but will try to avoid light.

Stored food product moths include Indian meal moth, Mediterranean flour moth, brown house moth and white-shouldered moth. Stored food product moths will eat stored food products in kitchens and pantries. Note that the white-shouldered moth will feed from both textiles and stored food products.

Treatment

- Launder infested clothes and/or dry them in temperatures hotter than 60°C. This will kill all life stages of moth. You can also heat curtains and other textiles, rugs, shoes, backpacks, stuffed animals, toys and similar objects by putting them in the tumble dryer at this medium-high temperature. These items do not have to be wet.
- All stages of moth will be killed if the infested objects are left in a freezer at 0°C for three days. Putting infested furniture outdoors during winter

may kill some moths, but there is no guarantee that you will kill all of them and the furniture may be damaged if moisture is present.

- Vacuum crevices around and under beds and sofas and in wardrobes that have carpet in them. If you're using a canister vacuum, immediately empty the contents into a plastic bag, seal and throw away, and clean the vacuum parts thoroughly. If you're using a vacuum with a bag, immediately remove the bag and seal in plastic for disposal or empty straight into the outside wheelie bin.
- Insecticides can be bought from shops or online to treat infested areas. Ensure affected carpets are thoroughly treated and rugs are treated on both sides.
- Insecticide fogging can be used, but please remember the gas will only kill off the adult moth and not eradicate the eggs in the moth's metamorphic life cycle.

Silverfish control

Silverfish are common in high-rise apartment blocks and offices where they thrive in dark, humid and damp conditions. They are mainly found around kitchens, bathrooms and basements, but they can survive in most environments. Their name derives from the silvery shimmer of their bodies and their fish-like movement when crawling.

If an infestation of silverfish is left untreated, then mass numbers can cause damage to wallpaper, books, paintings, pictures and anything containing starch or cellulose. They feed on cereals, moist wheat flour, starch in book bindings, paper on which there is glue or paste, mould and fungus.

They are fast moving and can travel throughout buildings. This pest is nocturnal but will also be active during the day in darkened areas.

Treatment

- Ensure your property has no damp issues.
- Improve ventilation.
- If you live in a block of flats, especially newly built flats, speak to the concierge or neighbours to understand the level of the infestation throughout the building.
- Spread diatomaceous earth powder where the silverfish are seen.
- Use an insecticide that leaves a residual layer behind which remains active for weeks after the treatment.
- Install dehumidifiers in damp areas in your home to stop silverfish returning.

Spider control

Spiders are found in almost every habitat, including our homes and workplaces. Many people suffer from arachnophobia, which is a fear of spiders. We are most likely to see spiders in late summer and autumn when they are at their largest and looking for a mate.

Some spiders, such as the false widow, can bite, leaving sore red marks. I have been bitten by a false widow several times when putting my hands underneath kitchen units, so always be careful if you're feeling around in darkened spaces.

Check for gaps around windows and door frames as these are likely points of entry for spiders. Also regularly check in loft spaces as these may be the arachnids' nesting site before they squeeze between light fittings to enter the living or working space.

Treatment

- Spiders can do good as they feed on other insect pests, so before rolling up the newspaper to kill a spider, where possible, catch it and release it back outside instead.
- If you really can't bear the thought of a spider in your home, some insecticides will eradicate arachnids.
- An old wives' tale of putting chestnuts around the home to prevent spiders coming in may be true.

Several of my customers swear by this method, even though there is no scientific proof that this will work.

Wasp control

Give wasps a wide berth because of their nasty sting, which the insect can repeat without causing any harm to itself. Some people, like myself, have a severe allergic reaction to the sting of a wasp, so it's always better to call in a professional if you find a nest in or around your home or business.

Most wasp species found in the UK build their nests in many locations – lofts, sheds, old rabbit and vole burrows, air bricks, cavity walls, chimneys or anywhere that is dry and undisturbed. They usually prefer higher locations but can be found at ground level. The nest will start off in the spring when the queen wakes from hibernating over the winter and will continue to grow until the following winter, when nearly all the nests die off due to lack of food as plants and flowers wither and other insect species die.

Treatment

- Always seek professional help with a wasp infestation for health and safety reasons. If you do attempt to get rid of the wasp nest yourself, wear a bee suit to protect yourself, especially in loft spaces where this pest may be nesting.

Woodworm control

Woodworm bore tiny holes of 1–1.5 mm in diameter in wooden items or structures. Live infestations show powder called frass (which is made up of wood particles bored out by the insect larvae) around the holes. Adult beetles are found in the summer months when they emerge from the wood to mate. The woodworm season usually runs from May to October.

The most common wood-boring beetle found in British buildings is the common furniture beetle, but the deathwatch beetle is also an indigenous British insect. Their larvae prefer sapwood and hardwoods, usually oak, which have partly decayed or are damp.

Wood-boring beetles can fly. Their flight is limited, but enough to get them through open windows like any other flying insect. The biggest risk of acquiring these pests is when you're purchasing old or second-hand furniture. Be sure to inspect these items carefully for any pinprick holes.

Treatment

- There are self-help insecticides that you can purchase online. These work well for small furniture items.
- I recommend getting a specialist to come and survey the site. Woodworm can be very

destructive to the supportive structure of some buildings.

CASE STUDY: OVERRUN BY ROACHES

I was called to a charity housing association block in Central London that was plagued by numerous pests. The property housed over 300 residences, and many of them were inhabited by cockroaches and mice. I entered one room that was believed to be the centre of the cockroach infestation and I had never seen anything like it.

The room was a box shape measuring about 4 m × 3 m. Upon opening the door, German cockroaches started crawling out into the corridor. More started dropping from the ceiling in the room. Thousands of cockroaches were on the walls, ceiling and crawling over the belongings in this room. I could barely see the cream-coloured walls due to how severe the infestation was. It reminded me of the scene in *Indiana Jones and the Temple of Doom* when all the insects start crawling over the characters when they are trapped in a chamber.

I returned to the property the following day to fog, wearing a hazmat suit with gloves and welly boots, and wearing a respirator. The building managers had turned off the fire alarm system and all people were cleared from the area for health and safety purposes. I stood at the door to the room and began fogging it with an insecticide. The cockroaches started falling from the walls and ceilings in panic. Some made their way out of the room trying to escape, and many crawled onto my suit and over my full-face respirator. I had my hood up on my suit, so no orifice could be breached. After 20 minutes of fogging, combined with a wet insecticide spray, 99% of the infestation was eradicated. The roaches that had made their way into belongings and under the occupants' clothing to avoid the treatment were eventually killed by us as we bagged and removed all the items from the room.

Owing to the severity of this one infestation, the problem had migrated through the building and encroached on an adjoining office. It took months of treating and controlling isolated outbreaks before we gained full control of the cockroach infestation.

Summary

Controlling insect pests is not always easy and trying to outsmart a stubborn infestation can cause you many sleepless nights. Following some of the simple steps we've discussed in this chapter and using the correct products, whether they be fogging agents, traps or insecticides, in the correct manner is key to gaining control quickly. Always be mindful of the health and safety of yourself and others, whether they be people in the same home or workspace as you, or your neighbours, when you're using these products, and if you're in any doubt as to whether you can control the infestation yourself, please call in an expert. I particularly recommend this course of action in the case of wasps.

Remaining calm and acting in a timely manner is essential when you're controlling an insect infestation. Always remember, speed is key.

SEVEN

Check

In this chapter, we will look at the fifth principle of the 6Cs: checking. We will examine the importance of this principle in both military and pest-control contexts so that any problem can be nipped in the bud rather than escalating. We need to remain vigilant because, put simply, what has happened once can always happen again.

First, let's look at checking from a military viewpoint.

Military check

After the Army I worked as a security engineering contractor for many years in the Middle East. I was on a task on Bagram Air Base in Afghanistan, about a 2-hour drive from Kabul. I was with my colleague Tony, who had completed a 22-year military career in the Army before working as a private contractor. We were overseeing the procurement and installation of ten-metre-high watch towers which would be installed on base or shipped to other camps around Afghanistan for over watch guard protection to help in the event of an attack on that camp. We didn't have a vehicle to move around the large air base, which housed over 40,000 personnel, so we got one from our company's Afghanistan-based driver, Amin, who was in Kabul. After arranging a pass to allow the vehicle on to the base, I met Amin, who had driven the white 4×4 from Kabul, at the crowded main-entry control point for the camp, where over 100 vehicles were leaving and arriving at the base to deliver and pick things up daily. After a quick vehicle safety search by the American military, the 4×4 was allowed in.

Tony and I used the vehicle for two weeks. Once we completed our work, we decided to head back to the capital to stay at the Baron Hotel next to the Kabul Airport and International Airfield (KAIA). The vehicle was required back in Kabul for the transportation and movement of other staff, and for the time being, we had no reason to return to the air base. Even though the road move may not have been the safest option as we weren't carrying any pistols, it was certainly the quickest with limited fixed-wing and helicopter flights for contractors. Besides, we were in an unmarked vehicle, and we planned to stay on the main open road which never had a great deal of traffic on it. It also had lots of panoramic views, so we could see things far in the distance. If there were any problems such as pop-up terrorist checkpoints, we would be able to turn back to safety.

We travelled on a Thursday afternoon because the following day, Friday, was the day of worship for the local people and there would be a crowd of other contractors in the hotel that night to speak with which was always nice. We set off on the Airport Road at 2pm with me driving. When we were 40 miles into our 70-mile trip, the fuel light pinged up red. Neither of us had thought to fill up the tank as the fuel gauge was still reading half full.

Even though it would be dangerous to break down in Kabul, we decided to plough on, believing we should have enough fuel to get us to the hotel. Tony kept in close communication with our team of four people back in the Afghan capital, where Amin, our driver, was ready to set off at a moment's notice with a jerry can of fuel if we did break down.

We reached the edge of Kabul and started driving on one of the most dangerous roads in Afghanistan, if not the world, due to the vast amount of vehicle-borne IEDs that had killed hundreds over the years. Tony forewarned the others of our location, which at that moment was next to Camp Phoenix, an American base stationed at the side of this road. These camps and convoys were targeted by terrorists, often when vehicles were slow-moving as they entered these compounds, so whenever we saw a convoy of military vehicles when driving through the city, we, like many other contractors, would hang back a good 200 m or so, for fear of them being targeted. On this occasion, we decided we would drive about a kilometre on this dangerous highway, and then go via the back route through a side gate of KAIA, across the safe military air base, then pass through the camp to exit out of another gate which was at the entrance to the Baron Hotel. Simple – or so we thought.

The vehicle began coughing and Tony got on the phone to our team in the hotel, asking them to make their way towards us. We turned right on to a side road which led to a small British base called Camp Souter. The vehicle's engine shut off just after we'd turned, so I free-wheeled the vehicle for as long as I could until we stopped about 500 m away from the gate of the British base, pulling over to the side of the road on a dirt verge.

Now it was a waiting game. Without any weapons to aid us, we were like sitting ducks if a hostile crowd formed. Immediately, children started to come around the vehicle, asking for water. We did have a couple of cases of water in the boot, but if we gave one out, more people would be drawn to us. Then some teenage boys and several men became curious and surrounded the vehicle. The doors were locked, but it wouldn't take much to smash a window. We had discussed jumping from the vehicle if that did happen and making a run for Camp Souter. I joked that I only had to outrun Tony, which made us laugh – nervously.

Ten minutes later, Amin arrived on site with another of our team, sixty-year-old Steve, who looked exactly like a lean Santa Claus. The Afghan people respected him as he was seen as an elder and this would help deescalate a difficult situation, if there was going to be one. Amin jumped from the vehicle with Santa. They got a jerry can of fuel and funnel from the boot and begun filling up our vehicle while Tony and I sat, somewhat relieved, selfishly not leaving the vehicle to help. I saw them both laugh while talking to the crowd and felt my nerves start to relax. Amin only put in enough fuel to get us away from the area as quickly as possible.

After several turns of the ignition, the vehicle fired into action. We waited for them both to get back in their vehicle, and then we all moved off together. It was the least we could do! We arrived at the side gates of the airport, showed our badges, and then drove through to the Baron Hotel.

That evening, as we sat down with a beer in the hotel bar, Tony and I talked about how lucky we had been to have escaped without any escalation of the situation. The funny thing was that neither of us had even once thought to check why the fuel gauge stayed on half-full the whole time we had been driving the vehicle. I blame Tony and he still blames me.

This anecdote stresses the importance of the fifth C – checking – in a potentially volatile military setting. Let's now see how this relates to the world of pest control.

Pest check

Keeping a constant watch for the return of a pest infestation that you have successfully treated once is crucially important. You also need to check in with those around you. It could be that you are doing everything to keep the pests at bay, but your next-door neighbours are not, and their pests are migrating over to your property.

Being vigilant is crucial to prevent another outbreak happening. Just as it did with Tony and me in the military story, complacency could land you in trouble and lead to serious unwanted results, so when it comes to pests in your home or workspace, keep regular checks ongoing.

Let's now have a look at how you can check that specific types of pests don't return to your property.

Mammal check

The methods you use to check a rodent infestation does not re-occur will depend on which rodent you have successfully controlled. The return of larger rodents such as **rats** and **squirrels** will be easy to identify as you will hear and recognise the tell-tale noises of the animal's footfall as it moves around, as well as loud gnawing on wooden joists.

Squirrels when nesting inside will tend to do so in loft spaces, so you will hear them above you first thing in the morning as they leave the property for their day in the great outdoors and in the evening when they return, as well as occasionally during the day if checking in on their young. To look for evidence of squirrels in the loft, you can shine a torch around the space after dark. You may see eyes reflecting back at you or hear them run away when opening the hatch.

Rats move around throughout the night in the cavities of the property, but if you have a basement and you are unsure if the rat problem has been resolved there, I recommend leaving some tempting bait out for the animal, such as chocolate spread or a breakfast bar. Even a piece of bread will do the trick. Rats are greedy animals and will feed on most things.

Droppings are a tell-tale sign that **mice** have returned. Mice will also feed from tempting bait such as chocolate spread, and this can be left on a piece of paper out of the way of pets and children, ideally under one of the white goods such as the fridge or behind the kitchen kickboards at the bottom of the units.

It is key to check for new entry holes that have been created by rodents. Their teeth can go through wood, plasterboard and thin aluminium sheeting. I have even seen evidence of a squirrel going through a lead plate on a church roof. If you do block the holes, use steel mesh or plating.

If you are confident in doing so, I recommend removing the kitchen kickboards and blocking the holes around all the incoming water pipes and electrical cables. Due to these gaps and holes essentially being there to allow entry to your property, I would advise stuffing the areas around the pipes or cables tightly with wire wool. The rodent will pull on the wire wool, which will slip through its teeth and cut its gums.

Foxes are particularly persistent mammal pests. Unfortunately, once one fox has been removed from an area, it is likely another will take up its place soon after, so be sure to check for tell-tale signs. Live sightings of foxes or spotting fox poo in your garden or on your land are the most common indications this animal has returned. You may find washing going missing from the line or items such as footwear being stolen from the back door.

Insect check

Ants. Check around all the areas you previously saw the ants. Most times, ants are attracted by sweet foods such as jams in the kitchen or pantry and it is likely they will have entered under doors or via gaps under the kitchen units. They may even enter through breaks in the brickwork pointing, air vents, poorly fitted windows or the footings of your home or workplace.

If the ants are black garden ants, trace them back to the nest. Search outside the property because garden ants, as the name suggests, make their nests in soil. Be mindful that these ants will return annually if you have not killed off the queen.

Bedbugs will usually be found anywhere people spend long periods of time resting. They are visible with the naked eye and prefer wooden or creased fabric nesting spots such as the flaps of a stapled headboard cover. Headboards are an ideal nesting spot because it is easy for the insect to travel to feed off the host, which is you, so I recommend you always remove the headboard when checking, as well as the mattress. Also check the edges of the bed's wooden slats and framework. Bedbugs tend not to nest around metal-framed beds.

Carpet beetles are obvious, as you will see these insects flying about. They are more prevalent during the spring and summer months when the adults can be found around windows as they are attracted to the sunlight.

I once had a customer who'd had six treatments done by a well-known pest-control company for carpet beetles around her penthouse flat in London. Two heat treatments and four wet insecticide treatments had not alleviated the problem, so she called me to investigate.

The problem had to be either in her clothing or external to the flat. It was the latter. The carpets in the communal corridor were crawling with woolly bears, the furry larvae of this insect, with the adults flying under the gap of her front door. Once the carpets were treated throughout the corridor, the problem was resolved.

Cockroaches can only survive a short number of days without water, so are likely to harbour close to a water source such as the condensation drip tray behind the fridge freezer. If you are checking for re-infestation, remember to check bathrooms, kitchens or water-tank rooms too. Use sticky pads to monitor this crawling insect.

Fleas. You will know if you have not managed to control a flea infestation as you will be bitten by them. If you have a pet, use a flea comb on their fur weekly at a minimum. You can buy sticky flea traps to identify fleas, but I advise you to use these traps for indication purposes only. They won't kill off the infestation as the many eggs and larvae of the insect that are waiting to complete their metamorphic cycle to adulthood will not be caught in the traps.

Flies are identified by a visual inspection check. Keep an eye out for larvae (maggots). Some infestations begin at the bottom of a waste disposal bin where the bin's inner bag has split, leaving the residue of rotting food, which is sometimes difficult to notice. This rotting organic matter acts as a perfect egg-laying spot for some fly species, so clean your bins thoroughly on a regular basis. If an infestation of flies is ongoing for several weeks, then searching for the source is key in solving the problem.

Fungus gnats. Inspect each plant you have in your property. Check its base, gently looking through the growing foliage to find signs of the clear or whitish fungus gnat larva with its shiny black head. If you find these larvae, quarantine the plant(s) from others for at least three to four weeks. This gives you plenty of time to spot weak newly emerged fliers, as well as establish temporary control methods before the gnats can spread.

Use sticky yellow traps to find flying adult gnats. Since it only takes one gnat to lay potentially hundreds of eggs, you need to keep the adults at bay. For an indoor plant, these sticky yellow card traps work well.

Moths. I advise you to use pheromone traps that can be placed around the property to attract the male moths, but these are only as an indicator to let you know you have a problem and give you an idea of where the infestation may be. For example, it may be in the wardrobe if lots of moths are on the trap in that location. Put these traps all around the home, especially in darkened wardrobes if carpeted or in the eaves.

Silverfish do not leave many tell-tale signs, so again, look out for their presence. Place sticky traps where the insect is likely to be, which is mainly in warmer, danker rooms such as bathrooms. The kitchen and boiler room are also favourite harbouring spots for these guys.

Wasps. Most of the time, wasps will be nesting in the roof spaces of homes and workplaces. When you're entering a confined, dark space to look for evidence of wasps, take care. I recommend you use a professional to check your property for these stinging pests. They will attend with all the correct safety equipment, such as a bee suit.

You can remove the nest once the wasps are killed off, but it's not necessary. No queen will ever occupy a pre-made nest, so you can leave a dead one where it is if it's in an unused space or too awkward to reach.

Woodworm. Check wood by covering all the previous holes with a wood filler. Even painting over them with a thick coat will fill the holes. If woodworm is still present, new holes will appear.

If you have woodworm on vertical surfaces or overhead, such as in the understairs cupboard, or in the floor panels above the basement or in the loft space, vacuum up all the previously created frass from the holes, and then place a piece of paper below where the infestation was. Check regularly to see if new frass is falling on to the paper. Be sure to bang on the holes, both before you vacuum them out prior to making new checks and when you are checking your paper, the vibrations shaking all the existing frass out.

CASE STUDY: A SAD DISCOVERY

During the COVID pandemic of 2020–21, the PestGone team and I were on a housing estate in southwest London looking at the external rat situation around the accommodation blocks. There was an ambulance on site, the paramedics removing a human body from one of the flats. The estate manager explained to us that the woman who had died had likely been lying in her flat for five or six months as no one had seen her in that timeframe. The pandemic had made it more difficult for neighbours to call in on each other, especially the elderly, to see how they were doing.

We had treated this property a year earlier for mice due to the customer leaving pet food down when the pet was not feeding.

We offered our services to check for any more pest activity and carry out any works to remove the carpets and carry out a biohazard clean where the woman had been lying. This resulted in us removing all the contents from the property due to her having no family, which was heart-breaking.

This case study highlights a different perspective on checking. If you haven't seen or heard from a neighbour in a while, then please do go and check in on them.

Summary

Continually checking for pests, even when you have treated the initial infestation, is important to ensure that the problem has been fully eradicated. If an infestation keeps rearing its head, then it is likely that the treatment has not been successful. There can be several reasons for this, including:

- The pesticides you have used are not toxic enough to kill off the pest.
- You didn't manage to treat all the areas where the pest is harbouring.
- External factors outside your property are drawing more pests to your location.

In this chapter, we have covered a number of ways by which you can check to make sure that your property is pest free today and remains so into the future.

EIGHT
Communicate

In this chapter, we will look at the sixth C: the vital principle of communicating with others. Clear communication is the key to solving any problems successfully, and it is no different when it comes to pests. Communicating with neighbours or the people you live with will help you get rid of your pest problem far more quickly and easily than if you go it alone.

I'm sure you can appreciate the need for clear communications in a military setting. Here's a story of communications getting somewhat lost in translation, leading to a potentially embarrassing situation for a very important guest.

Military communications

I was on an operation tour with the Army in 2006 in Kabul when the regimental sergeant major of my unit grabbed hold of me.

'Moseley,' he said, 'you are going to be having breakfast with a senior member of the royal family in half an hour.' When he named the royal in question, I was a little bit shocked by this news, but of course, I obliged. We were talking about a very important person, after all.

At that time, the sergeant major and I played football together for the regiment, so maybe that was why he picked me. Maybe it was because I was the senior electrician for the camp. Either way, at least I knew I was going to the front of the dining queue to get my food, which I was not too sad about.

When I went to pick my breakfast at the hot plate, I knew I had to be distinctly cautious of what I chose to eat. I couldn't have baked beans because I didn't want to have bean juice dripping down my chin while I was talking to Their Royal Highness. I was hungry, so I got a couple of poached eggs – well, they were supposed to be poached, but were rather more like two rubber squash balls bouncing on to my plate – some hash browns and a cup of tea.

I sat next to Danny, the interpreter for the British general who was head of NATO forces at the time, and we had a quick chat before our royal visitor walked into the dining hall. It became noticeably quiet, just a soft murmuring as Their Royal Highness came over with the general and sat opposite Danny and me, speaking to one of their aides in civilian clothing and indicating that just a coffee would suffice for their breakfast.

I said, 'Morning, Your Royal Highness,' to kick things off, but being a flamboyant character, Danny led us straight into conversation. He asked how things were with Their Royal Highness, who responded by enquiring what Danny's role was.

Their Royal Highness went on to tell a story of when they were in a particular country in the Middle East. A member of the country's royal family had had a child out of wedlock, which was highly frowned upon in the area's culture. The local interpreter, who was going to be translating for Their Royal Highness while they addressed a crowd and the TV cameras, had been told by his seniors to misinterpret what our royal visitor was saying. As a result, he translated their words in a way that mentioned the child born out of wedlock, even though Their Royal Highness had never uttered a word about this child.

Our royal visitor told us that they only found out about this later. As I'm sure you can imagine, they were not amused, telling us at the breakfast table, 'The only bastard that day was the bastard interpreter.' I tried not to choke out my hash brown, quickly coughing and managing to take a sip of tea to regain myself.

It was an amusing story, I thought – not that Their Royal Highness thought so – and I learned something more important that morning: that if somebody like this high-ranking royal, who seemed remarkably down to earth, is one of the people, maybe not running the country, but certainly at the top of the tree, then the UK is in good hands, because they came across as lovely and on the ball, which was refreshingly nice to see and something I did not expect.

Four weeks later, I was walking down the corridor to my accommodation when I saw people in suits through the open door to Danny's room. It was unusual to see people dressed in this manner as everyone generally wore military desert uniform, with the civilian personnel wearing cargo-type clothing, so people in suits suggested something important was going on. I guessed they were probably the special investigation branch of the military.

'Everything OK?' I asked the man holding the door to Danny's room open. Luckily, he was British, so he responded.

'I can't say, but the guy who was in this room won't be coming back.'

A few days later, I found out that the men in suits had confiscated Danny's computer and that he was found to be a spy who had been emailing sensitive information to Iranian officials locally. They found coded messages to an Iranian military attaché in Kabul reading, 'I am at your service'. A UK court found Danny guilty after a lengthy trial and he got ten years' imprisonment.

Years later, when I'd left the Army and was working as a contractor on military camps, pictures of Danny would come up in security briefings to explain how this person had got into the military, fooling everyone with his exuberant nature. I wanted to stand up and tell the people in the briefings that I knew Danny and he'd been a nice bloke who taught salsa, but I didn't want to be one of 'those people', so I kept quiet.

It was a strange time, now I think back on it. Me, having breakfast with one of the most famous royals in the world while sitting alongside an Iranian spy. It sounds like something that should be in an Ian Fleming novel.

Pest communication

I have added Communication to my pest principles due to a phone call I received recently. A young professional woman living in southeast London had had a problem with mice for several months and my team had successfully eradicated two infestations, but the problem kept returning every six to eight weeks in the form of tiny juvenile mice. I explained to the client that this would coincide with the gestation and weaning period of the species.

I asked the client if she could speak to her immediate neighbours as one of them owned a dog and may have been leaving pet food out. One of them might have had adult mice breeding and living in their property, and the juvenile mice were getting kicked out of this territory by the adult male as they became a threat to his harem of females and his home. Because of this, the young mice were migrating into my client's home.

However, the client wouldn't speak to any of the neighbours for fear of being frowned upon and seen as dirty because of her persistent mouse problem. She also believed it would be an insult to her neighbours to imply their property might be the root cause of the problem. She even refused to ask on the community phone messaging group if anyone else had recently started getting mice in their property.

I was baffled that someone would rather live with a constant stream of mice than discuss the issue with their neighbours. I grew up in Norfolk where people are, in my opinion, more social towards each other, and yes, they may know more of your business than you would like, but no one would want to keep up appearances or worry about the social hierarchy of the street while allowing a potentially harmful problem to get worse. Ironically, this mirrors the behaviour of rats. If one rat high up in the social hierarchy is eating something that is foul and disgusting, then the rats lower in the order will copy and eat the same foods to ensure they don't become outcasts.

As we're humans and not rats, we all need to get back to basics and communicate better when it comes to pests. If we are truly worried what the Joneses next door will say about a mouse or a rat that has entered the property, we should probably eject them from our social circle. Some years back, a row of terraced houses in Blackheath, southeast London, had a major rat problem, but not one person communicated with the others in the street. I eventually had to visit several houses and ask everyone living there to join a Facebook group. A week later, a broken sewer line under one of the affected houses was found to be the source of the infestation, which was successfully eradicated.

Because the people in this street had not communicated, they had suffered three years with rats, paying thousands of pounds to deal with the problem in their individual homes, not to mention two landlords losing their tenants who moved out immediately upon seeing the problem. This could have been resolved so much earlier, and more cheaply, if only people had spoken to one another.

If you take one thing away from this book, make it to communicate with your neighbours if you discover a pest. I am aware that some people have antisocial neighbours or those who clearly don't care about their property, and this is problematic, but overall, most people are more than happy to help. In an article, philanthropy advisor Jenny Santi says that it is scientifically proven people feel better during and after helping someone.[17]

Communicate with your neighbours, especially if they are old, vulnerable or have not been seen in a while.

Drop a card through the door. They may think you're being nosey, but at least you will have done your bit and you may just save someone's life. There will always be pests where there are humans, so it's better to work together to control them than try to do it alone. Communication between neighbours is also likely to lead to new friendships and a happy place to live with a real sense of community.

CASE STUDY: WOODWORM IN THE RAFTERS

A flat owner in an accommodation block of thirty flats in southwest London had woodworm in his kitchen units. He replaced the units, but the woodworm returned within weeks.

I was called to inspect the problem and explained that it was highly unlikely the woodworm originated in the customer's kitchen units. It was more likely that they had travelled down from the wooden roof framework as his flat was on the highest level. I checked the roof space, and sure enough, there were joists that had woodworm damage.

I advised the customer to get everyone who lived in the block together, either physically or by speaking over a social messaging service, to ascertain the true extent of the problem. By communicating in this way with his neighbours, my client found that four of the other top-floor flats had woodworm.

My client had lived with these beetles crawling around his flat for two years. Another pest-control company had advised him that the units could not be treated as the kitchen cupboards had a protective coating that insecticide wouldn't penetrate, and a heat treatment would damage the coating. That information was correct with regards to the treatments, but the key to solving the woodworm problem once and for all was ascertaining whether the root cause lay outside of the flat and communicating with the neighbours to gauge the extent of the infestation. If he had done that, this client would have saved £6k on new kitchen units.

Summary

The story at the beginning of this chapter highlights what happens when we don't communicate properly. In anything we do, clear communication is important to help us maintain productivity and build relationships.

The chapter's case study shows that speaking to others around us will help if we are dealing with pests. In this instance, remaining silent and allowing an infestation time to grow throughout the neighbourhood was the worst thing that my client could have done. Nipping unwanted visitors in the bud early is vitally important, so please talk to others in your property and/or locality if you are experiencing a pest problem so you can work together to combat it.

In my business, communication is vital to keeping my customers happy, so I ensure that I complete a report after each treatment or inspection so a paper trail is in place that the customer and I can refer to later. I either leave this with the customer or send it digitally later. If the communication channels between us are active and informative, then everyone knows what's going on with the pest problem, which is important to save the client time and money.

NINE

Commercial Pest-control Programmes

If you are a business owner with a commercial property, this chapter is for you. Without a pest-control maintenance programme in place, you run the risk of any pest infestation returning, even after you have treated it successfully. In this chapter, I will lead you through everything you need to know about a pest-control programme, including the cost, what is included and the all-important question: why you need one.

My business PestGone specialises in setting up pest-control maintenance programmes to protect some of London's well-known restaurants and buildings. From time to time, pests will enter a property, especially in a densely populated place such as London and even more so if that property houses a business that is serving food. The smells that restaurants, cafés, fast-food outlets, etc naturally emit will inevitably get pests such as mice, rats and cockroaches knocking at the door – or rather, inviting themselves in without knocking. As many buildings have cavity walls made from plasterboard or thin plastics in restaurant kitchens, it's no wonder that pests breach these and feast on the good stuff.

Having a regular pest-control maintenance programme in place, you will be able to nip these unwanted visitors in the bud early, as well as monitor for an onset of pest activity. Once you have established the programme, I recommend monthly inspection visits by a pest-control expert. This is especially important if your business serves food, but necessary even if food is simply stored or consumed on site. I can't tell you the number of times I have been into offices in the heart of London, hunting for mice that have been attracted in because members of staff have left chocolate in their drawers. No wonder mice and rats have got a taste for chocolate when it is so readily available due to it being left exposed to any passing pest.

Does your business require a pest-control programme?

A common question my team at PestGone gets asked is: 'I need a pest-control programme and want a certificate to prove it's in place, but I don't have any pests as my business isn't open yet, so what's the cost for a one-off visit?' Let's break this down to help you understand what you require from your pest-control programme. Below are the points we will cover:

- Do you get a certificate?
- Do you need a pest-control programme if you don't have pests?
- How much does it cost?
- What does the programme cover?
- Can you do your own pest control?
- Do you get a certificate? (Repeated on purpose, as this comes up time and time again.)

Let's answer each question in turn.

Do you get a certificate?

First and foremost, there is *not* a *Charlie and the Chocolate Factory* golden-ticket-style certificate which makes your business exempt from inspections and, worse, from being shut down. A pest-control programme is ongoing and tailor-made to your business to protect you, your staff and your customers from pests and their detritus. If a pest-control company arrives and issues a certificate of compliance after one visit, what's to stop pests infesting the very next day? A pest-control programme must be an ongoing process.

Do you need a pest-control programme if you don't have pests?

You don't have to have a programme run by an expert pest-control company, but if you are serving food to the public, you do need reliable, visible methods and records to show that you are controlling any and all pest infestations. Any infestations that occur must be logged for the relevant inspecting authorities.

If I had a pound for every time a customer has told me that when they opened a new establishment, they didn't have pests, but a month later, the property was infested, then I would be a wealthy chap. The reason pests are not present before your business opens is because there is nothing onsite to attract them or make them want to stay close by. As soon as they realise a constant supply of food is on offer, thanks to the scent of either the cooking smells coming from a restaurant kitchen or food left lying about by staff members, then pests will think they have hit the jackpot. It's then that an infestation grows – and a small problem turns into a seriously large one.

How much does it cost?

Any reputable pest-control company will give your business a free survey and advise on costs and the number of visits required, as every programme is different. What it needs to cover depends on a number of criteria, including:

- How big is the premises? If it's a seven-storey hotel, then the costs to keep it pest free will be higher than those of a small high-street coffee shop.
- What is the premises used for? For instance, a doctors' surgery may get the occasional mouse while a Michelin-starred restaurant serving thousands of plates of food a day and keeping the rear kitchen door open to allow ventilation is likely to attract a lot more pests.
- What is the location of the business? A bistro pub in the countryside may encounter the odd field mouse or flying insect throughout the year; compare this to a takeaway restaurant in the heart of London which will attract pests, such as house mice, rats, cockroaches, flies and foxes, in abundance due to the human traffic and food wastage.

To give you an example of cost, at the time of writing, a London restaurant that seats up to 200 people at a time will require eight to twelve inspections a year at a cost of £60–£100 per visit. If a property is prone to cockroaches, especially Oriental cockroaches, then the cost of the programme may increase due to the pest-control expert needing to make more visits to deal with a constant infestation coming from the drains or nearby buildings.

What does the programme cover?

A pest-control programme should at a minimum cover the main disease-spreading vermin such as mice, rats and cockroaches, depending on the location of the property. A maintenance folder is left onsite, containing the pest-control company's health and safety policy, environmental policy, COSHH information, risk assessment, method statement and monthly treatment reports. The treatment reports make up the paperwork an inspecting health officer is most likely to want to see, so you must ensure these are at the ready in the folder or on record somewhere. If these reports have been sent via email by the pest-control company, as PestGone does because digital data will rarely get lost in transmission, I would advise you to print them off and put them in the maintenance folder.

Can you do your own pest control?

The simple answer is yes! If you put in place visible methods and keep records to show that you are controlling all pest infestations when they occur, this is adequate. Any infestations that do occur must be logged for the relevant inspecting authorities.

Be aware if you choose this option that health authorities may scrutinise your methods if they see them to be inadequate and/or failing, and you must address them accordingly. Please don't hide the truth. If your premises gets a mouse, then log it. This will show the health inspector that you are carrying out the correct procedures and maintaining standards, which will get you a tick in the procedural box.

Legislation used in pest management in the UK

In the pest-control industry, we must comply with and adhere to several rules and regulations. Below, I have highlighted the key legislations, which can all be found on the Health And Safety Executive website (www.hse.gov.uk).

Health and safety legislation

Environmental Protection Act 1990. This act safeguards the public and the environment from pollution and waste, including noise pollution such as audible bird scarers. If pests cause a danger to human health in a business premises, prosecutions can and probably will be issued.

Health and Safety at Work Act 1974. This outlines the duties of employees, employers and the self-employed. In the context of pest control, this will cover pest-control technicians and a business's pest-management methods.

Management of Health and Safety at Work Regulations. These regulations fall under the act above and are mainly concerned with risk assessments.

Working at Height Regulations 2005 – regarding safety when working at heights, for example using a ladder.

Confined Spaces Regulations 1997 – proper training, a special permit and specific risk assessments are required to be able to work in a confined space.

Pesticides and food, and environmental legislation

Control of Substances Hazardous to Health (COSHH) 2002. These regulations fall under the Health and Safety at Work Act 1974. All pesticides, insecticides and rodenticides must display the active ingredient and other important safety information on the label. The first rule of COSHH is to substitute any chemical with a non-chemical method if possible, or if a chemical must be used, it must be controlled by law.

The **Food and Environment Protection Act (FEPA) 1985** aims to protect public health, animals, plants and the environment. This act makes sure information about all pesticides in the UK is available to the public. It defines responsibilities for pest-control officers (PCOs) and pest firms to use safe and efficient methods for pest control.

Control of Pesticides Regulation (COPR) 1986. This legislation regulates the approved pesticides for use by PCOs in the UK. Only certified technicians can apply professional products, which they must do in the safest way possible, taking all precautions and wearing the correct personal protective equipment. Pesticide waste is also regulated by the COPR.

Animal welfare legislation

Pests Act 1954. This act relates to rabbits as pests in the UK and regulates their control. It is the land occupier's responsibility to control the rabbits, not the landowner's.

Wild Mammals (Protection) Act 1996. This act makes it an offence for any person to mutilate, kick, beat, nail or otherwise impale, stab, burn, stone, crush, drown, drag or asphyxiate any wild animal. For fox control, only a single shot can be used to eradicate the animal.

Animal Welfare Act 2006. This defines responsibilities of animal owners and keepers to provide animals with a safe environment, diet and protection from harm. It also safeguards animals that are trapped, snared or held, as well as those being transported.

The **Wildlife and Countryside Act 1981** protects animals such as red squirrels, voles, bats, snakes, frogs, stag beetles and others. The act prohibits the use of bows and arrows, explosives or self-locking snares to control pests. It also prohibits the hard release of invasive species such as grey squirrel, mink and Canada geese if they are caught and so requires the animal to be humanely killed. The act does allow some birds to be legally controlled under general licence for a specific reason.

Public health legislation

The **Public Health Act 1961** grants powers to local authorities over premises infested by insects, parasites, rodents and birds.

The **Prevention of Damage by Pests Act 1949** defines responsibilities of local authorities to destroy mice and rats on their land. In some cases, the local authorities may force occupiers or landowners to act against rodent issues, as well as serve improvement or prohibition notices.

If you are a tenant in a property that has mice or rats, then it is the landlord's responsibility to treat the problem or pay for the treatment. Always check your tenancy agreement before signing because some landlords will put a clause in that will void this responsibility if the property is prone to mice and rats. Then it will be up to you as the tenant to arrange or pay for the treatment.

Food Safety Act 1990. All food that is sold must be safe for consumption, which involves keeping it free from contamination by pests. Environmental health officers may serve improvement and prohibition notices, fines and penalties to land or business owners if they don't comply with this act.

Food Hygiene Regulations 2013. All people who handle food commercially must be trained in the food-management system, Hazard Analysis and Critical Control Point. This includes making sure adequate pest-control procedures and methods are implemented with the food-management system in work.

Things to note

- There are a lot of regulations and hoops to jump through, but getting the right things in place early, regardless of how menial they may seem, is vital to keeping your business on track.
- Appoint someone in the business to be fully aware of the regulations and maintain standards. So many times, I have seen businesses where no one knows who is responsible for reporting pest issues. The employees point at each other and blame each other while the pest situation escalates out of control.

CASE STUDY: PIZZA PESTS

I was looking after a chain of fifteen pizza takeaways across London and I had highlighted concerns over many months about two of these premises. At one, rats from a damaged sewerage pipe in the walls that the management had refused to fix were gnawing through the soft plasterboard to gain entry to the cooking and serving part of the shop. At the second premises, there was a constant flow of mice because the shop's external bin shelter was situated immediately next to the rear kitchen door. This door was left open while the shop was operational from 5pm–11pm and the mice walked freely from the bin area into the kitchen. On one occasion, so did a rat. These animals were attracted by the smells of the pizzas cooking and the various toppings on offer.

Over six months of continuous reporting of this serious problem, I advised the managers that the rear doors of this takeaway premises must be closed. The chefs were not happy with the idea as the kitchen would become too hot, so I advised the company owners to install an air conditioning unit and a better extractor fan for the steam. They refused to do this due to costs, so the mice kept walking through the door every night.

Then inevitably, the health inspector did a spot check on the property. Finding mice droppings, they gave the company owners an improvement notice. The owners called me, demanding I meet them at the shop in question, which I did. I explained the situation clearly to them, pointing out the waste bins and puddles of fat that were only a metre or so from the open kitchen door. I showed them every report where I'd highlighted this problem and my advice to keep the door closed, but the company owners got angry, acting as if I had somehow caused the mice to come in. They still refused to install an air conditioning unit for their staff, so things carried on as normal and the back doors stayed wide open each evening.

Amazingly, the health inspector did give a health pass to the shop, but within days, more mice were entering. I found that they were feeding in the dough, so I decided to part ways with the chain. I would never want to work with company owners who refuse to protect their staff and their customers.

Summary

If you are a business owner with commercial premises, especially if your employees serve or store food, be sure to work with a professional pest-control company using the most up-to-date and proven practices to create an ongoing pest-control programme. This is important to nip any infestation in the bud before it can escalate into a problem that may even cost you your business. With the knowledge you have gained from this chapter, you can devise and manage your own pest-control programme, but I urge you to think carefully and seek professional advice if you're attempting to do this.

If you own a company serving food, always prioritise the health and safety of your staff and customers.

TEN

The Future Of Pests

Pests were here before us, and pests will be here long after we humans are gone. They are resilient blighters that I respect enormously. Take the bedbug, for instance. In the 1940s, we humans had nearly eliminated all bedbugs from the UK, but due to international travel and restrictions placed on effective pesticides, we are now suffering the worst bedbug crisis the UK has ever seen.

The pest-control industry will be here for as long as humans and pests live side by side. At the time of writing, the industry in the UK is worth around £790 million a year,[18] with a global annual worth of around £25 billion.[19] It's no wonder companies like Rentokil and Ecolab have seen huge profits and success for nearly a century. Recession and pandemic proof, the pest-control industry is buoyant and here to stay for the long term because pests will never be fully eradicated.

With the introduction of new technologies into the market every year, the industry is also getting safer as we reduce our reliance on old methods and toxins. An article by Luisa Reis-Castro says that scientists have even been modifying the genes of insects to limit their capabilities and reduce their life span.[20]

In addition to gene modification, the 2015 Science and Technology Select Committee Report on Genetically Modified Insects states that researchers have discovered ways to use an insect's biology against it.[21] Scientists are studying and implementing growth regulators to stop insects such as bedbugs from reaching their adult stage. This disrupts their growth cycle, which removes their ability to reproduce, stopping an infestation developing. An article in the *New Scientist* tells us that rodenticides have also evolved with some now giving female rats an early onset menopause and affecting the sperm of male rats. Over time, this will reduce numbers by making rats infertile, which is seen by some as a better choice than killing them. This was trialled in New York in 2013 and a study suggested populations were reduced by half.[22]

This method is useful to control the rats in the sewers who will be unable to reproduce but will still live out their normal lifespan. The downside to this treatment is if you have rats nesting in your home or workspace.

Then, it would not be the viable option to control the infestation in the building; only eradicating the animals early and/or blocking them out is key to preventing restless nights and worry. The other concern is that the longer these treatments are used, the more chance there is that the animals will become resistant to them over time.

Giving back

As a business, PestGone must kill animals. None of my team enjoy doing this, and neither do I, but it's an unpleasant necessity. As a result, we wanted to find ways to give back and help the environment.

We contacted several charities and organisations but struggled to fully connect on a personal level with those we spoke to until we were introduced to B1G1 (www.b1g1.com). This organisation's passion and professionalism inspired us to want to do more, so we joined it, knowing that all the money we donated would have the desired effect. Some of my team are, like me, ex-military and had been stationed in Africa and Brunei, so it felt right to help projects in these areas.

For every customer my business serves, a tree is planted in the rainforests of Indonesia and a Kenyan farmer will receive the means to continue bee farming, generating an income for them and their family as well as helping to rejuvenate the globally declining bee populations.

Not only do we give to help bees thrive, but we give to protect and replenish the rainforests that act as the world's lungs and are home for so many amazing species.

It may seem ironic that a pest-control company that eradicates insect and animal infestations should decide to protect insects and animals around the world. All the species we treat for here in the UK are pests, some of which carry and spread disease, such as rats and fleas. Placid insects like bees are becoming endangered.

Bees are the world's biggest pollinators. According to bees and pollinators advisor Alison Benjamin, a third of the world's food production relies on bees.[23] The Beatlie School's blog post on World Bee Day in 2021 said that bees pollinate (a process vital to crops growing) 75% of leading global crops, including oilseed rape, apples, soft fruits, beans, tomatoes and strawberries.[24] Bees are vital for maintaining the balance between living organisms such as plants and animals as well as their environment. It's estimated that if we lost all the wild bees in the UK, it would cost the farmers around £1.8 billion a year to manually pollinate their crops.[25] Remember, if we lose bees, there is no plan bee!

Bees also help reduce pollution. The honey and wax that they make is tested for toxins to give scientists a better idea of the air pollution levels in some cities such as Rome. The BBC's *Newsround* mentioned in 2018 that historically, bees have been used in pollution-reducing research in New York, Tokyo and London.[26]

These are just some of the many reasons we all need to do our best to protect this gentle insect. As an ethical company, PestGone will not treat for bees and will only attempt to relocate them if the bees are causing health and safety issues to the customer. At the time of writing, amazingly, bees are still not a protected species worldwide.

How you can help:

- Plant bee-friendly plants like heathers, daisies and red clover on balconies, terraces and in gardens.
- Leave sections of your garden wild and let the grass grow long. This gives bees a place to shelter.
- Create a bee hotel for your garden's bee population.[27]
- Leave a small dish outside with a few pebbles and shallow water for thirsty bees to drink from.
- Provide a sugary drink for an exhausted bee.
- Avoid using pesticides as they are harmful to bees.
- Buy honey and other hive products from your nearest beekeeper.

Some people think bees are aggressive, but nothing could be further from the truth. Bees are a placid insect that will not sting unless they're provoked or need to protect their queen. If you leave bees alone, they will happily go about their buzziness. The bee knows that it will die from stinging you, so will only do so as a last resort, unlike a wasp, which has multiple stings. Wasps are aggressive flying arseholes; bees are friendly, non-aggressive and essential to our environment.

If you're unsure how to distinguish a wasp from a bee, here are a few hints to help you:

- Most bees will have some furry hairs, while wasps have none you can see with the naked eye.
- Wasps are bright yellow and black, while most bees, with the exception of the bumble bee, are orange/gingery and black.
- Bees are not interested in your sweets or booze because flowers are their food of choice. Wasps will be a nuisance around food waste, sugary drinks and alcohol.
- Bees are gentle natured and reluctant to sting, as opposed to wasps who are aggressive and will sting repeatedly.

CASE STUDY: BEE STOPS

Most of us now know that bees are on a huge decline with so many species listed as endangered. We know of their importance, but at the time of writing, little is being done to protect them. It's down to us as individuals to do what we can to help our bees.

People often find bees in their offices or on the street and help them by giving them sugary fluids so they can go on their merry way energised. Taking this a step further, The Netherlands introduced something in its cities called 'bee stops', which are little gardens containing grass and wildflowers on the roofs of bus stops. The aim is to encourage pollination and contribute towards biodiversity, absorption of rainwater, reduction of the urban heat island effect, capturing particulates from the air and making the city look a far greener and nicer place. The UK city of Leicester has also adopted the idea of bee stops.

One day, I would love to walk down Oxford Street in central London and see bee stops all around me. I would certainly devote my own time and my company's time to help with installing and maintaining these.

Summary

As long as there are humans on the planet, there will be pests making our lives difficult. As we have seen in this chapter, research is ongoing into safer ways to control pests, but they are persistent creatures. The future of pest control is buoyant.

While we face an ongoing battle against pests, protecting all our native wildlife is important, whether that be bees, birds, bats, red squirrels or any of the other endangered species. All these creatures are fighting to survive in our ever-changing climate and environment, so where possible, we need to do our bit in whatever small way to help them.

Conclusion

Pest control is not the sexiest of subjects to read about or be a part of, but it is a necessity in our villages, towns and cities to protect our properties, health and wellbeing. I may seem to make light of a serious subject in some of the stories in this book, but all of them, from both a military and a pest-control point of view, were somewhat worrying experiences at the time. It's good to be able to look back and laugh.

I would like to thank you for taking the time to read through the 6C pest principles:

1. Confirm
2. Clean and clear
3. Cordon and contain
4. Control

5. Check
6. Communicate

If there is one thing I would like you to take away from this book it is to be able to look back on certain situations and try and find the funny, lighter side of them – and, of course, knowing how to deal with your own personal pest problem. I have worked in the homes and offices of thousands of customers from all walks of life, dealing with all manner of pests, and the one thing I have found in general is that, especially in London, these people are tired and stressed, feeling the world is against them. Some even believe their current pest problem has targeted them on purpose. Following the 6Cs principles will help you successfully rid your property of your pest and get your life back on track.

I want to end the book with something meaningful to me. I remember a film called *The Edge* starring Anthony Hopkins and Alec Baldwin.[28] In that movie, a plane crashes in the wilderness. The two lead characters survive the crash, but then have the task of making it back to civilisation while a wild bear stalks them.

During one scene, Anthony Hopkins's character has the insight that most people lost in the wilderness would die because they didn't do the one thing they needed to do to save their lives. When Baldwin's character asks what that one thing is, the reply is 'Thinking'.

Thinking has always resonated with me. Often, people give up or get someone else to do basic things for them, which to a certain extent is fine as it keeps small businesses like mine afloat. One client paid me £200 in the form of four crisp pink £50 notes because he didn't know how to change four spotlights in his home. I could comfortably reach the spotlights by standing on the bathroom floor and he provided me with the four replacement lamps. Within fewer than two minutes, the spots had been replaced. He was surprised at how simple it was. All that the man had needed to do was go online and watch a self-help tutorial video for ninety seconds and he would have saved himself £200. I didn't even ask for any money; he offered it willingly in the hope I would change the lights.

Changing four lightbulbs requires a different set of skills to surviving in the wilderness, but the common thread between the two is not thinking. Most individuals fail at whatever it is they want or need to do because they don't think things through first. I have procrastinated a lot while writing this book, always letting something else take priority and not focusing on one task at a time. This shifting from task to task was me not thinking, so to avoid the shame of failure, I had to stop, think and prioritise.

When it comes to pests, I recommend that you do the same. Stop creating stress for yourself, think about the best way to get rid of your pest problem in your home or workspace, and prioritise this over most other things. Pests can be harmful to your health and destructive to your property, and if one has managed to infiltrate the building's fabric, more will too.

Stop, think, prioritise, and the bedbugs won't bite, the rat's race will be run and the mice will cease to play, whatever the cat is doing. Good luck!

Notes

1. A Augustyn, 'Effects and significance of the Black Death' (Britannica, 2022), www.britannica.com/event/Black-Death/Effects-and-significance, accessed October 2022
2. 'International tourist arrivals reach 1.4 billion two years ahead of forecasts' (UNWTO, 21 January 2019), www.unwto.org/global/press-release/2019-01-21/international-tourist-arrivals-reach-14-billion-two-years-ahead-forecasts, accessed October 2022; 'Shipping fact: Shipping and world trade: driving prosperity' (ICS, no date), www.ics-shipping.org/shipping-fact/shipping-and-world-trade-driving-prosperity, accessed October 2022
3. BBC, 'Elephants have "best sense of smell"', *Newsround* (23 July 2014), www.bbc.co.uk/newsround/28445461, accessed November 2022
4. University of Würzburg, 'New research finally explains why spider silk is so incredibly tough', *SciTechDaily* (September 2019), https://scitechdaily.com/new-research-finally-explains-why-spider-silk-is-so-incredibly-tough, accessed October 2022

5. 'Timeline of Rentokil history' (Rentokil, no date), www.rentokil.co.uk/timeline, accessed October 2022
6. J Bellis, 'New List: Top 50 best restaurants in London 2022' (Eating in London, April 2022), https://eatinginlondon.co.uk/top-50-best-restaurants-in-london, accessed October 2022
7. W Disney, *Disneyland* (television programme, October 1954), available at Sam's Disney Diary, 'The Disneyland Story' (YouTube, 2020), www.youtube.com/watch?v=LtdO-Gf8mWU, accessed October 2022
8. Aquahoya1, 'Mickey Mouse creator Walt Disney suffered phobia of mice' (Flip the Movie Script, 2014), https://flipthemoviescript.com/mickey-mouse-creator-walt-disney-suffered-phobia-of-mice, accessed October 2022
9. T Dow (director), J Sullivan (writer), 'Fatal extraction', *Only Fools and Horses* (episode aired BBC1, 25 December 1993), www.imdb.com/title/tt0666546, accessed October 2022
10. 'What to do if you see someone selling or using glue traps' (PETAUK, no date), www.peta.org.uk/action/see-someone-selling-using-glue-traps, accessed October 2022
11. Wildlife and Countryside Act (UK Public General Acts, 1981), www.legislation.gov.uk/ukpga/1981/69, accessed October 2022
12. 'Defence of Rorke's Drift' (National Army Museum, no date), www.nam.ac.uk/explore/defence-rorkes-drift, accessed October 2022

13. PPC94, 'Rat behaviour in rodent control' (BCPA, 2019), https://bpca.org.uk/News-and-Blog/rat-behaviour-in-rodent-control/222905, accessed October 2022
14. Natural England, 'Foxes, moles and mink: How to protect your property from damage' (Gov.UK, 2016), www.gov.uk/guidance/foxes-moles-and-mink-how-to-protect-your-property-from-damage, accessed October 2022
15. Natural England Species Information Note SIN004, 'The red fox in rural areas' (Bounty Pest Control, no date), www.bountypestcontrol.co.uk/pest_control/file/Red%20Foxes%20in%20Rural%20Areas.pdf, accessed November 2022
16. Section 1(1) (a), Wildlife and Countryside Act (1981)
17. J Santi, 'The secret to happiness is helping others', *Time* (no date), https://time.com/collection/guide-to-happiness/4070299/secret-to-happiness, accessed November 2022
18. 'Pest control in the UK: Market size 2011–2029' (IBIS World, 2022), www.ibisworld.com/united-kingdom/market-size/pest-control, accessed November 2022
19. 'Pest control market by pest type' (Markets and Markets, no date), www.marketsandmarkets.com/Market-Reports/pest-control-market-144665518.html, accessed October 2022
20. L Reis-Castro, 'Genetically modified insects as a public health tool: Discussing the different bio-objectification within genetic strategies' (National Library of Medicine, December

2012), www.ncbi.nlm.nih.gov/pmc/articles/PMC3541591, accessed October 2022
21. Science and Technology Committee, 'Genetically modified insects: oral and written evidence' (Parliament UK, November 2015), www.parliament.uk/globalassets/documents/lords-committees/science-technology/GMInsects/GMInsectsevidence.pdf, accessed November 2022
22. A Klein, 'Menopause-causing bait is curbing rat populations in New York', *New Scientist* (May 2017), www.newscientist.com/article/2130114-menopause-causing-bait-is-curbing-rat-populations-in-new-york, accessed November 2022
23. A Benjamin, 'Why are bees important? You asked Google – here's the answer', *The Guardian* (17 June 2015), www.theguardian.com/commentisfree/2015/jun/17/why-are-bees-important, accessed October 2022
24. 'World Bee Day' (Beatlie School, May 2021), https://beatlieschool.westlothian.org.uk/article/69545/World-Bee-Day, accessed November 2022
25. H Vickers, 'Why are bees important? And how you can help them', (Woodland Trust, 17 July 2018), www.woodlandtrust.org.uk/blog/2018/07/why-are-bees-important-and-how-you-can-help-them, accessed November 2022
26. 'Bees get busy keeping pollution in check', *Newsround* (BBC, July 2018), www.bbc.co.uk/newsround/45003800, accessed October 2022

27. 'Build a bee hotel: How to make a bee house' (Friends of the Earth, no date), https://friendsoftheearth.uk/bees/make-a-bee-house, accessed November 2022
28. L Tamahori (director), D Mamet (writer), *The Edge* (1997), www.imdb.com/title/tt0119051, accessed October 2022

Acknowledgements

I would like to thank everyone who has, in some way, helped to create and been a part of this book. I would also like to thank the countless people who have supported my business in the past and those who continue to do so.

Thank you to all of the current and past PestGone staff, including Sean, Stephen, Sammy, Dave, Jamie, Jack, Daniel, Barry, Tessa and George. To anyone I served with in the military, with a special mention to 42 Field Squadron Royal Engineers and Power Troop RE – Allied Rapid Reaction Corps. To the people I worked with in the security sector after my military career, including Richie, Tony, Mark, Pete D, Pete H, Pete N, Jim, Harriet, Andy, Paul, Nikki, Alan, Steve W, Adele, Ravi, Simon, Des and Jake.

I would like to thank Dent Global and the business connections I have made through this company, including

Dave, Emma, Megan, Alex, Sophie and Elyse. Thank you to the team at Heropreneurs, including Amy, Jenny and Rob, with special mention to Jon, for your guidance over the years. I really appreciate all your input.

I would like to thank all the people who have helped me in the pest-control industry. Michael and Des of Combat Pest Control, Gary of West London Pest Control, Jim of Protex Pest Control and all the staff at Killgerm Chemicals, especially Melvin and Richard – you have all supported and continue to support me and the business.

Thank you to all the people I had the pleasure to work with on *The Apprentice*, and thank you to Lord Sugar for the opportunity. To my fellow candidates, Shannon, Sohail, Megan, Avi, Emma, Simba, Rochelle, Greg, Dani, Joe, Victoria, Reece, Shazia, Kevin, Denisha, Brad and Marnie. I wish you all the luck and business success for the future and thank you for making my time on the show such a fun and memorable experience.

I would like to thank all the customers PestGone has had the pleasure of working with and continues to serve. I'm sure you hope never to see me or the team in a professional sense again, but if you do require our services, you know where we are.

Finally, I want to remember those I have had the pleasure of serving and working with who are no longer with us: Sam, Shaun, Lew, Lee, Mike and Steve. Thank you.

The Author

Mark Moseley joined the British Army at the age of sixteen, serving nine years with the Royal Engineers. During his service, he carried out multiple operational tours including Afghanistan, Northern Ireland and the DRC. Upon leaving the Army, Mark backpacked around the world for eighteen months, travelling through every continent except Antarctica. Returning to the UK, Mark worked in the security sector operating in hostile environments in the Middle East and Africa for five years before speaking with a close military friend who encouraged him into the pest-control industry.

Mark is the founder of PestGone Environmental Ltd, a London-based pest-control company. Since starting in the industry in 2016, Mark has featured in a number of national newspapers and on a high-profile BBC1 business show called *The Apprentice*. He has given talks about his journey into pest control and the advantages former military personnel can bring to the workplace to senior facility management companies and to government representatives in Whitehall. His thousands of customers include members of the royal family, the rich and the famous, and he has partnered with government authorities across London.

In his spare time, Mark enjoys visiting new countries, mountaineering, flying, painting and watching his football team, Everton. You can connect with Mark and find out more about his business via:

www.pestgone.co.uk

www.linkedin.com/in/mark-moseley-0434b3123

PestGone Environmental Ltd

@Pestgonee

@Pestgoneenvironmental

Printed in Great Britain
by Amazon

18738498R00109